KEEPING FOOD SAFE

KEEPING FOOD SAFE

The complete guide to safeguarding your family's health while handling, preparing, preserving, freezing, and storing food at home

HASSELL BRADLEY
and
CAROLE SUNDBERG

Doubleday & Company, Inc.
Garden City, New York
1975

Library of Congress Cataloging in Publication Data
Bradley, Hassell, 1930–
  Keeping food safe.
     Bibliography
     Includes index.
  1. Food handling. 2. Food—Preservation. I. Sundberg, Carole, 1943–   joint author. II. Title.
     TX601.B73   641.3
     ISBN 0-385-05124-7
Library of Congress Catalog Card Number 74-15375

COPYRIGHT © 1975 BY HASSELL BRADLEY AND CAROLE SUNDBERG
ALL RIGHTS RESERVED
PRINTED IN THE UNITED STATES OF AMERICA
FIRST EDITION

Dedication...

To my parents, Frank and Kathryn Grimes, for the heritage which makes pursuing an idea a joy; to my daughter, Margaret, for her ideas and for keeping the home fires burning; to my son, Bill, for his enthusiastic interest and art work; and especially to my husband, Joe, for his constant devotion, help, and encouragement.

... Hassell Bradley

Dedication...

To my parents, Maybel and Leo Kadinger, for exemplifying hard work and honesty, and especially for their continuing sacrifice and love.

... Carole Sundberg

Acknowledgments

We wish to thank the following individuals and groups for their contributions, with a special thanks to Editha Quaintance, who gave unlimited time and energy beyond the call of duty.

Jean Anne Vincent, Editor, Special Projects, Doubleday & Company, Inc., whose support and encouragement have inspired us from the beginning; the United States Department of Agriculture for the help and interest they have shown in the book (especially Nancy Harvey Steorts, Special Assistant to the Secretary of Agriculture for Consumer Affairs, and Carl Sorenson, Information Officer, Agricultural Marketing Service, Southwest Region); Sentinel Newspapers, Denver, Colorado metropolitan area; Dick Hilker, Executive Managing Editor, Sentinel Newspapers; Colorado Extension Service; Colorado State University Library; Public Service Company of Colorado (Home Service Department); Public Health Service . . . Region 8 (especially Ralph C. Barnes, Regional Program Director for Disease Control, and the Public Information Department); Food and Drug Administration . . . Region 8 (especially Fred L. Lofsvold, Regional Director, Helen Keaveny, Consumer Affairs Specialist, Grace Paavola, Consumer Affairs Technician, and Denzil Inman, Regional Food Consultant); Dr. S. M. Morrison, Department of Microbiology, Colorado State University; Mrs. Anna Katherine Jernigan, Iowa State Department

ACKNOWLEDGMENTS

of Health; Marian Johnson, Association of Home Appliance Manufacturers; Alice I. Doherty, Western Region, General Electric Consumer's Institute; Ronald S. LaMotte, Colorado Outward Bound School; Jeanne Varnell; Nancy Porter, Jeff Varnell, and Neil Brown.

Contents

	Acknowledgments	*vii*
	List of Tables	*xi*
	Introduction	*xv*
1:	Misery à la Carte	*1*
2:	The Life-Saving Rules	*19*
3:	The Spoilers	*33*
4:	Problem Foods and Safeguards	*53*
5:	Clean and Healthy Kitchens	*95*
6:	We Are Bugged!	*115*
7:	Putting Up Food	*139*
8:	Stocking the Larder	*169*
9:	Investigation and Control of Food	*223*
	Bibliography	*245*
	Index	*259*

List of Tables

1. Food Poisonings and Infections
   page 18

2. Recommended Internal Temperatures for Meat and Poultry Cooking
   pages 31–32

3. Critical Temperatures for Safe Food Handling
   page 32

4. Guide for Storing Fruits and Vegetables
   pages 68–69

5. Common Household Pests
   page 138

6. Temperature of Food for Control of Bacteria
   page 168

7. Freezer Storage
   pages 220–21

8. Refrigerator Storage Chart
   page 221

9. Time Needed to Thaw Meat in the Refrigerator
   page 222

*Not to us, O Lord, not to us,
but to thy name give glory,
for the sake of thy steadfast
love and thy faithfulness!*

                    Psalms 115:1

Introduction

On August 14, 1973, the United States Department of Agriculture (USDA) and the Department of Health, Education and Welfare (HEW) held a joint press conference to launch a governmental broadside attack on *Salmonella* and other foodborne illnesses.

Dr. Charles C. Edwards, HEW's Assistant Secretary for Health, has since emphasized that improper handling of food at the retail level and in the home is a major source of *Salmonella* illnesses.

In December 1969 the Food Safety Panel of the White House Conference on Food, Nutrition and Health identified the problem of microbiological contamination—food illness—as a major American health concern. The Panel recommended wider surveillance of foodborne disease and a development of new testing methods to detect them. The panel reported:

> Not only do we need public awareness of the extent of foodborne illness, but also a mechanism is essential to get the practicing physician and individual citizen to participate in the reporting system.

No one knows precisely how many cases of food illness occur in the United States, but conservative estimates place the number between 2,000,000 and 10,000,000, and this is annually. The National Academy of Sciences estimates that *Salmonella,* only one of the food illnesses, costs Americans $3 million each year, plus an indeterminate amount of pain and discomfort (13).

Most of these cases and similar illnesses could be prevented. It is the rare American, however, who even wonders if the two-day-old leftover might be dangerous. Few of us are aware that the more serious forms of spoilage or infection rarely reveal themselves through odor, taste, or appearance.

The U. S. Public Health Service has also launched a vigorous campaign to cut down the toll of foodborne illnesses. Public Health Service doctors emphasize that while we cannot defend ourselves against all food illnesses by personal care and hygiene, we can reduce our risk considerably with a little knowledge and effort.

The main purpose of this book is to provide that knowledge about food safety. Is the food I am putting on my family's table tonight safe from disease and poisons? What are the necessary precautions I must take to keep our food safe? How do pests contaminate food, and what can I do about them? Do I have to discard all moldy foods? What should I know when I stock the larder? How can I apply the life-saving rules when I preserve foods? What can I do to protect the wholesomeness and safety of the meat, fish, and poultry I buy? Which three spots in my kitchen spread the most bacteria? What do I do if the freezer stops? What food laws protect us, and who enforces them?

The term "food illness" is used throughout the book as a general term for any disease caused by eating food. Technically, food poisoning refers to an illness caused by a poison present in the food when it is consumed.

Food infection is an illness caused by invasion, growth, and damage to human tissues by pathogenic organisms carried by food.

The main thrust of this book is to enable the average consumer, who has limited knowledge of microbiology to understand the dangers which are constantly present in his environment as related to foods.

Certain technical principles have been given in the first part of this book, so that the life-saving rules can be better understood and applied.

This is not a "scare" book. The food we consume in America

INTRODUCTION

is the safest in the world, but there are dangers. This is why we should know and apply the life-saving rules and principles outlined in this book.

This book will also help the consumer to understand carelessness in the kitchen may well have been responsible for that last siege of "twenty-four-hour flu."

This is not the usual type of book written about food and the preparation of meals. Up until now there has been no single source of information about preparing, serving, storing, and preserving foods with an emphasis on sanitation and disease-prevention measures. Individual bulletins, which tell exactly how to avoid food illness, have been available from the USDA, HEW, and FDA for years. However, few consumers know about or take the time to order these publications.

One dependable source of information about the care and handling of food and proper sanitation measures in the kitchen is definitely needed. This book attempts to fill that need.

KEEPING FOOD SAFE

1

Misery à la Carte

It was 9 P.M., June 29, 1973, somewhere in the tropical waters east of Miami. The luxury liner *Skyward* steamed at top speed toward port, its Caribbean cruise cut short. Inside the cabins most of the 900 passengers and crew were in their bunks, miserable with diarrhea, fever, and dizziness.

On the mainland, the Coast Guard was making hurried arrangements for an emergency airlift to supply the 600 passengers and 300 crew members with paregoric for their diarrhea.

The next day in Miami, passengers and crew members were placed in quarantine, examined, and treated. After everyone had calmed down, doctors from the National Center for Disease Control in Atlanta were able to pinpoint the ailment as caused by *Shigella flexneri,* a rather acute form of food illness (4).

No single food could be blamed for all the suffering. However, a definite link with the sickness was established with two shrimp dishes. A water test also showed a greater risk of illness for the passengers and crew members who had drunk more than two glasses of water a day.

The nightmarish incident was just one more in the long annals of food illnesses which have plagued mankind ever since the first cave dweller left the dinosaur meat out in the sun.

Whether it's just a queasy stomach or respiratory failure caused by dreaded botulism, food illnesses mean one thing: someone has been careless. Food infections and food poisonings are sneaky attackers, and they can strike down any one of us if we don't take simple precautionary measures. Yet, millions of us do not take these precautions. Food illnesses are not maladies that happen only to other people. Most of us have several cases during our lifetimes. Usually we dismiss each of these illnesses as "just another bout with the twenty-four-hour flu."

U. S. Public Health Service officials say millions of cases of "flu" each year are actually undiagnosed cases of *Salmonella* and staphylococcal poisonings.

The experts, who include scientists with government agencies, the food industry, and the Department of Agriculture, all agree there are three basic ways we can prevent food poisoning and infection in our kitchens.

These preventative measures are:

(1) Improve personal hygiene.
(2) Use only clean work surfaces, dishes, utensils, and equipment.
(3) Take simple precautions in storing, preparing, and cooking food.

The misery we serve up for ourselves can be divided into categories, each with its own symptoms, peculiarities, and means of control.

Botulism

Some toxins produced in our favorite foods by microorganisms, such as the one produced by *Clostridium botulinum,* are so potent even a tiny amount can kill.

In Bedford Village, New York, two persons ate commercially prepared canned soup unheated from the can. Both became gravely ill, and one died. At Bakersfield, California, three months later, seven adults at a family meal tasted or ate chili sauce made from home-canned peppers. One died, and six others became ill, one mildly, two seriously, and three critically. In Philadelphia within the same time span, three people became ill after eating home-canned green peppers, and one died.

In each case the foods contained botulism toxin produced by the bacterium *Clostridium botulinum,* the deadliest food poison known to man.

Controlled experiments at the University of Michigan have shown that one trillionth part of a gram of pure botulism toxin is enough to kill. In fact, botulism toxin is so deadly it has been considered as a prospective bacteriological warfare weapon (16).

Botulism was first described as an official disease in southern Germany in the early nineteenth century after a study of more than 200 cases of sausage poisonings. (The word "botulism" comes from the Latin word for sausage: botulus.)

Between 1925 and 1972 there were only five deaths attributed to botulism poisoning from commercially canned products. However, during the same time, some 700 deaths were traced to botulism poisoning caused by improper home canning (15).

In this country inadequately home-canned foods are most often the cause of botulism. The main causes in Europe are preserved meats and fish. Of the canned foods, those most often responsible for botulism have been string beans, sweet corn, beets, asparagus, spinach, and chard, but other kinds of food have also been responsible for outbreaks. In general, the low- and medium-acid canned foods are most often responsible, but instances of poisoning from acid foods, such as tomatoes, apricots, pears, and peaches have occurred. These more acid foods had been grossly underprocessed, thus allowing the growth of other microorganisms to aid growth and toxin production by *Clostridium botulinum*.

Meats, fish, seafood, milk and milk products have also been responsible for outbreaks. Some recent cases of type E botulism in the United States have been caused by smoked fish which had not been properly refrigerated. An outbreak from canned tuna resulted from contamination of the tuna through defective seams in the can.

Botulism is the number one enemy we face when we can foods at home. Ironically, home-canned foods are symbolic today of that extra measure of tender-loving care, of a willingness to put forth extra effort in the kitchen. It is wise to cash in on savings by preserving when fruits and vegetables are plentiful and relatively inexpensive, but savings mean little in the face of tragedy.

Clostridium botulinum is abundant in cultivated and forest soils, bottom sediments of our streams, lakes, and coastal waters. It is often present in the intestinal tracts of fish and mammals and the gills and viscera of crabs and other shellfish.

Botulism spores, then, are found throughout the environment and are harmless. However, in the proper environment, and when not destroyed by heat, the spores divide and produce poisonous toxins. *Clostridium botulinum* is an anaerobic bacterium, which means it does not require oxygen and grows better in its absence.

Onset of the disease is usually within 12 to 36 hours, and it generally lasts three to six days. The earliest symptoms are an acute digestive disturbance followed by nausea and vomiting and possibly diarrhea, together with fatigue, dizziness, and headache. Later there is constipation. Double vision and difficulty in speaking and swallowing may occur. Patients may complain of dryness of their mouths and constriction of their throats. Their tongues may be swollen and coated. Body temperatures can be normal or subnormal. Involuntary muscles become paralyzed, and the paralysis spreads to the respiratory tract and heart. Death usually results from respiratory failure.

The fatality rate is high in this country (about 65 percent). It doesn't have to be fatal, however, if it is diagnosed in time and treated promptly. Supplies of antitoxin against the three main types of botulism poisoning known to affect humans are stockpiled at the U. S. Public Health Service Center for Disease Control in Atlanta.

This silent stalker is at our elbows in our kitchens far more often than most of us dream. The spores can survive in food being canned, processed, or preserved. Some foods are far more hospitable to the spores, of course, and when these are inadequately processed or preserved, a tragic set of circumstances can be set in motion.

But let's say we have been careful to follow correct canning or processing directions. We still cannot dismiss the possibility of botulism growing in our food and kitchens. Certain environmental conditions after processing will permit the germination of spores as well as growth and toxin production by the organism. Therefore, we still run the risk of death or at least severe illness by not heating canned foods sufficiently before they are eaten.

CONTROL MEASURES

The safeguards against botulism are little more than common sense, based on a knowledge of why and how foods spoil. Yet how many of us are really familiar with this area of microbiology? These life-saving rules are simple:

(1) Follow approved canning and preserving measures issued by the United States Department of Agriculture.
(2) Reject gassy (swollen) or otherwise spoiled canned foods.
(3) Refuse to taste a doubtful food.
(4) Avoid foods that have been cooked, held, and not well reheated.
(5) Boil home-canned foods for 20 minutes in an open kettle.
(6) Avoid these foods which have been thawed and held at room temperature: meats, poultry, seafood, fish, vegetables, milk, and milk products.

Recently new methods of smoking fish and meats have become popular. To prevent botulism from occurring in these smoked foods these steps should be followed:

(1) Strict food sanitation maintained throughout handling and smoking.
(2) During smoking or afterwards the foods must be heated for 30 minutes until the coldest parts reach 180° F.
(3) After smoking, the foods must be frozen immediately and kept frozen.

Staphylococcal—"Ptomaine" Poisoning

At a church dinner more than half of those who had eaten chicken (about 100 persons) were sick within six hours. In other cases, 17 people aboard ship became sick after eating a mixed salad, and 180 people became ill at a banquet after eating sliced meat.

Investigation into the church dinner revealed the chickens had been cooked the day before and immediately refrigerated. However, the next morning they were reheated and cut into quarters with a butcher's meat saw. The chickens were not refrigerated from 10 A.M. until 5 P.M., and the cook who had prepared the chicken had numerous small cuts and abrasions on his hands.

The shipboard illness was caused by a cook who mixed the salad with his hands in spite of the fact that he had several minor cuts on two fingers. In the case of the banquet, the meat had been sliced with a machine on a block. Both the meat-slicing machine and the block had been heavily contaminated.

The culprit in these food illnesses was the *Staphylococcus* organism. It has caused food illnesses throughout the ages, but the bacterium was first identified by Louis Pasteur in 1880.

In the year 1970 alone, 4,699 persons were known to have had "Staph." This was 19.8 percent of all foodborne illnesses which were reported. However, as the U. S. Public Health Service is quick to point out, figures like this represent only the tip of the iceberg (5).

The actual number of cases is not known because most people do not recognize food illnesses as such.

"Staph" cases usually are not reported at all unless the outbreak is significantly widespread. However, a large percentage of all food illnesses are *Staphylococcus* poisoning, and most of us come down with the illness a number of times during our lives (7).

Once it is understood where staphylococci come from, the control measures are obvious. The simplest sanitation and hygienic steps prevent most cases. What a shame so few people recognize the dangers in ignoring these simple measures.

Staphylococci are present on the outer surface areas of almost all animals. Most of us carry them on our skins. The nasal passages of many persons are laden with them and are a common cause of sinus infections. Boils and infected wounds can also be potent sources.

Of the many kinds of food which have caused "Staph" illnesses, custard and cream-filled bakery goods (which have been undercooked, contaminated after cooking and/or allowed to stand at room temperature), and ham, tongue, and poultry have caused the most outbreaks. More offenders are other meats and meat products, milk and milk products, fish and fish products, cream sauces, salads, puddings, pies, and salad dressings. Growth and toxin production of staphylococci may also take place in food kept warm for long periods under heating lamps, in steam tables, ovens, and food-vending machines.

The most common human symptoms are salivation, nausea, vomiting, retching, abdominal cramping of varying severity and diarrhea. (Sound familiar?) Blood and mucus may be in the stools in severe cases. Headache, muscular cramping, sweating, chills, prostration, weak pulse, shock, and shallow respiration may occur. Usually a normal or subnormal body temperature is experienced instead of fever. The duration of these miseries is brief, usually only a day or two. Most of the time recovery is uneventful and complete. No wonder the average victim thinks it's "just another bout with the flu." The death rate is extremely low. No treatment is given except in severe infections when saline solutions are given to restore the salt balance and to counteract dehydration.

We all differ in our susceptibility to "Staph," so of a group of people eating food containing toxin, some might get very sick, and a few lucky ones might be affected only mildly. The incubation period (the time between the actual eating of the food and the appearance of the first symptoms) is usually two or three hours, but it can range up to seven hours.

In order for an outbreak to occur, *Staphylococcus aureus,* a bacterium whose toxin is fairly resistant to heat, must be growing in a food which is a good culture medium. The food must provide the ideal conditions for toxin production by the cocci. Too few of us ever think of food as a culture medium for growing organisms.

The temperature of the food must be favorable to the growth of the cocci (between 50° and 115° F.) and enough time must be allowed for the production of the toxin.

CONTROL MEASURES

In order to prevent "Staph" illnesses we need to follow common-sense sanitary measures. In general terms, growth of the cocci must be prevented, and any surviving cocci should be killed. Perhaps the most important control measure for staphylococcal poisoning is to keep hot foods above 140° F. and cold foods at or below 40° F. Toxin is destroyed by boiling several hours or heating the food in a pressure cooker at 240° F. for 30 minutes.

The best practical safeguards are to use ingredients completely free of contamination, such as pasteurized rather than raw milk, and by keeping people who have colds, boils, and infected cuts away from foods.

Growth of the cocci can be stopped by refrigeration of foods and, in some instances, by an adjustment to a more acid state. Some foods can be pasteurized in our own kitchens to kill the cocci before they are exposed to room temperature.

Clostridium Perfringens Food Poisoning

One day almost everyone at an exclusive girls' school was sick. The suspected meal of roast beef and gravy had been cooked the day before and set to cool in open trays, without refrigeration, for 22 hours.

In 1970, *Clostridium perfringens* accounted for almost 30 percent of all victims and 15 percent of all outbreaks of food illnesses (5).

The spores have been found in samples of most raw foods and in soil, sewage, and animal droppings. Most foods commonly involved are meats that have been cooked, allowed to cool slowly, then held for some time before being eaten. Fish paste and cold chicken are common offenders.

Clostridium perfringens is a spore-forming bacterium that grows in the absence of oxygen. (Spores can withstand the temperatures we use in cooking most foods.) Surviving spores continue to grow, particularly in cooked meats, gravies, and meat dishes which are not refrigerated properly.

The symptoms, which usually appear within 10 to 15 hours, are severe abdominal pain, gas, and diarrhea, usually without nausea, fever, or vomiting. Recovery is rapid.

CONTROL MEASURES

The means of controlling *Clostridium* are deceptively simple. Adequate refrigeration will control the growth of the bacteria and spores. Yet how often we overlook this simple precaution, thinking there is little need for worry in our spotless kitchens, in a world that relies on miracle drugs.

Yet the statistics, representing only a fraction of the actual cases, show a need for positive action against this bacteria in our kitchens.

Salmonella Infection

In July 1970 a cook in a Baltimore nursing home prepared a meal for the residents. Within a few days after having eaten the meal, 25 patients died of *Salmonella* infection.

Increasingly, *Salmonella* is in the news. In 1963 there were 18,000 cases reported. By 1972 the figure had reached 22,151 cases. Thousands upon thousands of cases go undiagnosed and unreported.

Health officials say the true number is probably closer to two million cases a year, with a large percentage of those cases diagnosed as "flu." In fact, it is estimated that 1 percent of the entire population has *Salmonella* each year.

In August 1973 the U. S. Departments of Agriculture and Health, Education and Welfare launched a broadside attack against *Salmonella*. The campaign involves consumer education aimed at correcting careless food-handling practices in the home and food service establishments. In addition, federal-state-industry programs were initiated to control and eliminate *Salmonella* throughout the food chain.

However, Agriculture officials are the first to admit that *Salmonella* cannot be totally eliminated from the food chain, because it exists in a great percentage of the animals and birds consumed as food.

Salmonella organisms are often present in poultry, even when frozen. They may also be present in eggs, especially in those eggs which are frozen, dried, or cracked. But poultry and eggs are not the only culprits. Almost any raw meat may be contaminated.

All those pet turtles the dime stores used to sell were prime sources. So are the baby chicks still sold at Easter. Other extremely potent sources are ourselves and other human beings. All of our pets should be considered potential carriers. Much of the animal and bird life around us can be contaminated with any of a large number of different types of *Salmonella*.

Changing food habits and food processing advances have probably added to the rising number of cases. We eat more frozen, prepared foods. We eat more candy, ice cream, cake mixes, and other convenience foods. Most contain dried eggs. The organisms in the eggs may be dormant in a product when it is sold. Later, however, when the consumer mixes it with liquid, warms it, and lets it stand at room temperature, the organism is free to grow and reproduce.

Considerable attention is now being given to shell eggs and to liquid, frozen, and dried eggs as sources of *Salmonella*.

Large-scale handling of foods by commissaries or institutions tends to increase the spread of the illness. Food-vending machines and precooked foods also add to the risk.

The foods which have been found to be the most common causes of *Salmonella* outbreaks are meat, poultry, and their products, especially if these foods are held unrefrigerated for long periods.

Fresh meats may carry *Salmonella* bacteria that caused disease in the slaughtered animals; handlers may also contaminate the meat. Products such as meat pies, hash, sausages, cured meats (ham, bacon, and tongue), sandwiches, chili, etc., are often allowed to stand at room temperature, permitting the growth of salmonellae.

Poultry and its dressing, seafood and its products should not give trouble if handled and cooked properly. Foods made with raw eggs carry a potential risk of *Salmonella*. Pastries filled with cream or custard, baked Alaska, and eggnog can be delectable timebombs.

Incubation periods can be as short as 5 hours or as long as 72 hours. Individuals also differ in their susceptibility. Principal symptoms are nausea, vomiting, fever, abdominal pain, and diarrhea. These can be preceded by headaches and chills. Other symptoms are watery, greenish, foul-smelling stools, prostration, muscular weakness, faintness, usually moderate fever, restlessness, twitching, and drowsiness. The mortality rate is less than 1 percent.

The severity and duration also vary with the amount eaten. Intensity can vary from a mild stomache ache and diarrhea to death. Usually symptoms last for 2 or 3 days, followed by an uncomplicated recovery, but they can linger for weeks or months. Up to 5 percent of the victims become organism carriers.

To become ill, a victim must have eaten food which contained the organism in considerable numbers. The food must have been a good culture medium, and the temperatures of the food had to have been held between 45° and 115° F., long enough for the organisms to reproduce sufficiently. When food of this sort is eaten, all the conditions have been met for a good-sized serving of misery.

CONTROL MEASURES

Care and cleanliness in food handling and preparation are all-important. How often we are quick to criticize a restaurant or cafeteria when it is not clean, but how frequently we overlook the same conditions in our own homes. Anyone who handles food should be healthy and clean, and food should only be prepared in clean surroundings and with clean utensils.

A beautiful, juicy raw steak can contaminate a platter with live salmonellae on the way to the patio and the barbecue. If that platter is reused after the steak is cooked without first washing it with soap and water, cross-contamination will occur, and eating the steak will be dangerous. Anything raw meat comes in contact with should be washed with hot water and soap or detergent.

Leftovers often support the growth of salmonellae. Therefore, leftovers should be placed in the refrigerator, and reheated thoroughly before being served again. Canned foods which have been recontaminated after opening and then held at warm temperatures can also cause the illness.

Shigellosis

Although the passengers and crew of the *Skyward* luxury liner can tell us all about shigellosis, few of us have ever heard of it. Actually, the disease is bacillary dysentery, and is an acute infection of the bowel caused by organisms of the *Shigella* group and is characterized by bloody diarrhea. Shigellosis is a human disease and the bacterium can only get into foods or water through human recontamination.

The incubation period is from one to four days, but frequently diagnosis is wrong, because suspect food is not around a week or so later to be examined.

Shigella infections are usually spread by contaminated water, defective plumbing, or food contaminated by flies or unwashed hands.

From 1951 to 1960, 108 outbreaks of shigellosis were reported. The number of cases ranged from 6 to 1,400 per outbreak.

The true incidence is greatly underreported (16).

Viral Infections

Today scientists are learning that some of the viral diseases can also be transmitted through food. Viruses are microorganisms even smaller than bacteria. The viral infections which can be transmitted through food are infectious hepatitis, poliomyelitis, and virus of the Coxsackie and ECHO groups, plus a few others (7).

INFECTIOUS HEPATITIS

We risk infectious hepatitis when we eat raw shellfish (especially oysters and clams) which have been harvested from sewage-contaminated water. When other foods, such as milk and potato salad, are involved, the source is usually either contaminated water or someone who has had the disease.

It can be controlled by cooking all shellfish, pasteurizing milk, using safe water, and practicing the life-saving rules (15).

POLIOMYELITIS

One of the greatest medical triumphs of all time has been the development and use of the Salk and Sabin polio vaccines. The disease has received vast amounts of publicity. Yet few of us are aware of recent research which proves polio can be transmitted through food, primarily raw milk. Personal cleanliness of those handling milk, pasteurization of milk, and good public health practices will keep it from occurring (15).

COXSACKIE AND ECHO VIRAL INFECTIONS

The role of these viruses in foodborne illness is not really known. However, they are viewed with suspicion, according to the U. S. Department of Health, Education and Welfare. More research is being conducted in the field all the time (15).

However, we will be safe from all foodborne illness if we practice the life-saving rules and observe the salient points discussed in the following chapters.

Trichinosis

Let's say we have browned some pork chops which are 12 mm (or about half an inch) thick and cooked them at a braising (low) temperature for 25 minutes. Are they safe?

No. The internal temperature of the meat was not high enough to kill *Trichinella spiralis,* a parasitic roundworm which may have been in the pork.

Most human trichinosis is caused by eating raw or incompletely cooked pork containing the encysted larvae. The larvae are released into the intestinal tract during digestion. They then invade the mucous membranes of the first parts of the small intestine where they develop into adults. The fertilized females give birth to thousands of larvae, which then travel through the blood vessels and lymphatics to skeletal muscle tissues, where they encyst.

It may take four or five days or even several weeks to begin to feel the effects of trichinosis after eating infected pork. Symptoms may include nausea, vomiting, diarrhea, profuse sweating, prostration, thirst, weakness, edema of eyes, labored breathing, chills, and skin lesions. The symptoms may last for days. Later symptoms, caused by the migration of the newborn larvae to the muscles and their encystment, are muscular soreness and swelling. A severe case could result in death.

CONTROL MEASURES

In order to be sure we are safe, it is necessary to make sure that every part of the meat reaches at least 170° F.

Pork can also be treated by quick freezing or storage at 5° F. or lower for not less than 30 days or −20° F. for 12 days. Some of the time-honored methods of processing sausage or pork products are adequate salting, drying, and smoking. Refrigeration will also control trichinosis.

However, only the United States Department of Agriculture's formula should be used. Directions specify one part of salt per 30 parts of meat and holding in a drying room for more than 20 days at 45° F. or above. Smoking should be for 40 hours or more at 80° F. or above, followed by 10 days or more in the drying room at 45° F. or above.

Poisoning by Chemicals

Unfortunately, there are other ways we can poison ourselves when we place food on our tables.

Although the information is readily available in various government publications and bulletins, few of us know that we can make ourselves ill by storing acid foods in certain types of vessels.

Let's say we are planning an après-ski gathering tomorrow after a day on the slopes. Just before we leave the city we pack a decorative old copper pitcher. In the evening we can mix a fruit punch, store it in the pitcher, and serve it when we return.

But that punch can turn into a deadly potion. The acids in the fruit juices and the copper will react to produce a poison which could have very serious effects on anyone who drinks it. Vessels made of tin, silver, alloys, silver-plate-on-copper, and those which are galvanized (zinc-coated) are also dangerous.

Illness caused by chemicals is fairly common, unfortunately. Usually symptoms appear within a short time. The culprits make up a deadly rogue's gallery and include mercury, arsenic, antimony, cadmium, lead, and zinc.

Consider the insecticide sodium fluoride, which has been accidentally added to food instead of baking powder, dry milk, or starch. The illustrations could be endless.

We can also add poison to our diets by forgetting to wash fruits before we eat them; thus we eat arsenic residues from fruit sprays. Although these residues are usually present in harmless amounts, how much better it is to wash them off first, so not even a small amount can be consumed.

Sometimes suffering victims think they have been poisoned by a food when a chemical is actually the culprit. For instance, methyl chloride poisoning can occur from a leaking mechanical refrigerator.

To further complicate the situation, almost any substance is toxic if given in big enough doses. Multiple vitamins can be poisonous if taken in large quantities. Common table salt, mistaken for sugar and mixed in infant formulas, once caused the deaths of several babies in a hospital nursery.

Another source of toxic material can be the glazed pottery on many shelves.

The ceramic glaze put on dinnerware is a thin, glassy coating, generally containing silica, which is applied to pottery by dipping, spraying, or brushing. It is then fired at extremely high temperatures. After fusion the glaze looks like a pane of glass.

Glazes seal the surfaces of pottery and make them more resistant to wear. Some pottery materials absorb moisture or liquids unless they are glazed. The glaze helps prevent the absorption of organic matter into the foods or liquids and helps prevent bacteria from gathering on the surface of the dish.

Often, lead or cadmium and other toxic metals are used in the glaze. These substances do often pose hazards if the glaze has not

KEEPING FOOD SAFE

been properly mixed, fired, or applied. These dangerous metals can be released by high-acid foods and vinegar-containing foods, such as sauerkraut, tomatoes, and tomato products.

In such a case, the metals are released into the food or beverage if it stands in the container for long periods.

The Food and Drug Administration now monitors the production of ceramic dinnerware in the United States to make sure it does not constitute a hazard. However, the FDA has no way of monitoring ceramic ware produced in earlier years or in other countries.

Since there is no easy way to identify whether glazed dinnerware contains high levels of toxic metals, we can stay on the safe side by not storing foods or beverages in such containers for long periods, such as overnight. Daily use of the dinnerware for serving is not harmful.

Poisonous Plants and Animals

In August 1973 a Michigan man was vacationing in Green Ridge Campground at Shadow Mountain National Recreation Area in Colorado. He picked and cooked several of the poisonous *Amanita muscaria* mushrooms nearby and ate them. He took a nap, then awoke several hours later vomiting and hallucinating.

It takes an expert to determine which mushrooms are safe and which are not. The ranger who helped the victim get to a helicopter ambulance repeatedly warns vacationers that summer rains make mushrooms sprout in abundance in the mountains each year and that all of them should be avoided, no matter how attractive they are.

Other gastrointestinal disturbances or even death may result from eating young fava beans or even from smelling the blossoms of the plant. Snakeroot poisoning results from drinking milk that comes from cows which have fed on this plant or a similar weed. Even the familiar rhubarb plant has greens which can cause oxalic acid poisoning.

Many common shrubs used to beautify and landscape our lawns are extremely poisonous if any of the plant is eaten or even tasted. So many of the plants we are all familiar with are potentially toxic. A good rule of thumb is to never taste any unknown plants and to teach children to observe the same rule.

There are several general rules we can use to avoid poisonous plants and animals:

(1) Never eat any fish which doesn't have scales or which is an unusual shape, unless we know it is safe to eat.
(2) Red is a danger signal in the vegetable family.
(3) A good rule of thumb is to avoid any vegetable or fruit which is divided into five parts. There are a few exceptions to this rule, but we will be safer if we remember this simple caution.
(4) Avoid any fungus growth completely. A safe mushroom is exceedingly difficult for the novice to tell from a poisonous one.
(5) Avoid plants which have a milky sap and which have a bitter or soapy taste.
(6) Berries may be divided into three categories:
   a. White berries are usually nonedible.
   b. Blue berries are always edible.
   c. Red berries may or may not be edible.

Ocean mussels and clams during certain seasons of the year contain a poisonous alkaloid, apparently from plankton consumed by shellfish.

Animals and plants ordinarily considered safe to eat may become sources of dangerous food illness. For instance, our polluted waters may yield fish contaminated with mercury.

Yet in spite of the potential poisons which exist around us, mankind has turned his edible environment into a more friendly one.

The decades and centuries to come will see many changes in the attitudes we have toward the foods we eat. It will become increasingly important for us to be aware of the potential hazards which exist side by side with us.

TABLE 1

FOOD POISONINGS AND INFECTIONS

Illness	What Causes It	Incubation Time	Symptoms	Duration & Mortality		Foods Usually Involved
Botulism	Toxin-*C. botulinum*	12-36+ (hours)	Vomiting, diarrhea, fatigue, dizziness, headache. Dry skin, mouth, throat. Constipation, no fever, paralysis of muscles, double vision, respiratory failure.	1-10 days	65%	Low- or medium-acid canned foods: meats, sausage, fish & other types of seafood.
Staphylococcus poisoning	Toxin-*Staphylococcus aureus*	3 (1-6) (hours)	Vomiting, abdominal cramping, diarrhea. Low temperature. Sweating. Generally attributed to other causes.	1-2 days	Very low	Custard- or cream-filled baked goods, ham, tongue, poultry—their dressing & gravy, head cheese, meat sandwiches, salads, cream sauces, cakes, dairy products, etc.
Clostridium perfringens poisoning	Toxin-*C. perfringens*	10-12 (hours)	Nausea without vomiting, abdominal pains, diarrhea.	1 day	None reported	Inadequately cooled cooked meats, poultry, fish.
Salmonellosis	Infection-*Salmonella*	12-24 (7-30) (hours)	Vomiting, abdominal pain, diarrhea—sudden onset. Usually fever. Chills, prostration, severe headache.	2-3 days	1%	Meat products, warmed-up leftovers, salads, meat pies, hash, sausage.
Trichinosis	Infestation-*Trichinella spiralis*	48 (hours)	Vomiting, diarrhea, sweating, colic, loss of appetite. Muscular pains later on.	Weeks to months	1-30%	Raw or poorly cooked pork or pork products.

2

The Life-Saving Rules

Fire was tamed, anthropologists tell us, in 500,000 B.C. It may have been centuries later when a cave dweller dropped a piece of raw meat into the flames. Irritated, he probably raked it out with a stick, tasted it . . . and marveled.

The way we cook today is light years away from those first crude attempts at cooking. However, few of us are aware we have many careless habits which can cause foodborne illnesses . . . a major American health problem today.

Interestingly, one of the reasons early man's life span may have been so short was his constant poor health caused by vitamin deficiencies and foodborne illnesses.

It is appalling how little we actually know about the food we put in our mouths. We have been saturated with information about calories, proteins, malnutrition, and the "Fat American." Yet we have heard very little about the living organisms which inhabit our foods or how to control their growth.

The rules for controlling these organisms (and thus most food illnesses) are elementary. However, the failure to observe them is responsible for more misery than all other mistakes we make in handling food in our homes.

The three most important requirements for safe food handling are constant attention to rules of personal hygiene, food tempera-

KEEPING FOOD SAFE

tures, and sanitation in the kitchen. Any food preparation should always begin with clean hands, clean utensils, and clean work areas.

If we have staphylococcal organisms present in or on our bodies (and most of us do) and we merely breathe onto a pudding or have a little skin infection and get some of the "Staph" organisms into the pudding, it is contaminated. They will not grow and become dangerous, however, if we chill the pudding and serve it cold.

But if we allow that pudding to remain at room temperature, the organisms will multiply and can easily make us very ill. Unfortunately, we cannot tell if the pudding is contaminated or not until we eat it, because the "Staph" toxin does not change the appearance or the taste of the food.

Modern life, not unlike life in the caves, has complications which compound the problems. Even if we have never experienced it, we can sympathize with a young homemaker, dirty diaper in one hand, a crying baby in her arms, who answers the telephone to learn her husband is bringing the boss home for dinner.

It wouldn't surprise us if the young mother forgot to wash her hands before she pulled a hen packed in a plastic bag from the freezer. Maybe she used the same knife to cut open the plastic bag that she used to cut the chicken, then forgot to wash the knife before she sliced the bread. Perhaps she stuffed the hen before it was completely thawed, then cheated on the cooking time.

Any combination of factors like these can be a potential cause of food illness.

Because all raw poultry, meat, and fish should be considered as a source of *Salmonella* organisms, and because the young mother used the same knife, unwashed between chores, to cut poultry and then slice the bread, she risked cross-contamination from a raw meat product to a ready-to-eat product (6). Furthermore, the undercooked dressing posed a natural breeding ground for the *Salmonella* organisms in the hen. And finally, her unwashed hands spread bacteria over everything she touched.

"But," some scoffers may say, "mankind has made it this far. Why get all excited about foodborne illnesses?"

The answer to that argument is, obviously, all of us haven't

THE LIFE-SAVING RULES

made it this far without being the victim of food illnesses. Unless we know the rules for safe food handling, we are not safe and secure when we sit down to eat in our own homes. We get sick. We suffer. Some of us die from food illnesses.

Most of us are totally unfamiliar with the world of disease-producing organisms. Fortunately, most of the time we are able to live side by side with these organisms; we have built up an immunity. Over the years mankind has learned the undeniable importance of personal hygiene.

But millions of Americans do not know how to carry these simple hygienic measures a step further in their kitchens as they cook and serve food.

It is in these very foods that we eat and serve to our families that disease organisms are often given a perfect environment in which they can multiply many times over. In the face of overwhelming numbers of the organisms, we cannot help but fall victim to disease.

Many of our mothers and grandmothers observed hygienic measures routinely and rigidly. Some were firm believers in preventative steps before problems had a chance to occur. (Of course, they were not always successful, but at least they tried.)

But we are spoiled. We have our gleaming refrigerators, our freezers, our efficient ranges. Our doctors can give us any number of miracle drugs if we encounter infections and poisonings. So we think we are safe from the diseases our ancestors worried about . . . and died from.

However, we have become so complacent that we have forgotten what it is we should be concerned with, and most certainly we have forgotten the preventative measures some of our forebears observed when they worked with food.

Certain rules should be observed in our kitchens all the time to eliminate the bouts with food poisoning and/or infection which we often ignorantly call the "twenty-four-hour flu."

Remember how grandmother gathered the green beans in her apron in the garden, brought them in, drew a bucket of water and rinsed the beans again and again before she cooked them? The practice is still as important today.

Fruits and vegetables should be washed under lots of running water or through several changes of water. If the latter method is used, the produce should be lifted out of each water change so the dirt doesn't resettle on the food. This habit is particularly important, because garden soil contains the types of illness-causing bacteria that are the most difficult to destroy.

And what about that raw beef and chicken we learned is a prime source of *Salmonella?* When we work with these foods we should never touch anything else until we wash our hands thoroughly with hot water and soap. We should do all we can to lessen the possibility of contaminating serving utensils or cooked, ready-to-eat foods.

Meat or poultry which is thawing in the refrigerator should not come in contact with other foods or utensils. Blood from the raw meat can easily and quickly contaminate other foods.

Although cooking kills *Salmonella* "germs," the same platters and utensils should never be used for the raw meat and then for the cooked portion before first washing the dish with detergent and hot water. Ignoring this simple rule can result in recontamination of the cooked meats.

It is even better to use separate cutting boards for raw and cooked meat. If you have only one board and you have used it for cutting raw meat, thoroughly wash the board in hot water and detergent before using it for any other food.

It is also extremely important for us to learn to estimate the proper cooking time so that meat or poultry is fully cooked and served immediately when it is done.

A thick cut takes longer to cook than a thin cut of the same weight. A thick cut of meat should be farther away from the heat than a thinner cut to assure proper cooking when broiling.

Meat with an outside layer of fat requires more cooking time than meat with little or no fat cover. Boneless roasts take a longer time per pound to cook than roasts with bones.

All meats, poultry, and fish should be cooked at low to moderate temperatures. This provides maximum tenderness and juiciness, and yields more meat because of reduced shrinkage. More importantly, the center of the product is well cooked without the outside being overdone.

When meat or poultry is cooked in steam or liquid, the cooking liquid should be kept just below the boiling point to make sure the product will be cooked thoroughly.

Using a meat thermometer to measure internal temperature is the best way to test the doneness of a meat roast or roasted poultry. Use a thermometer that shows temperatures in degrees as well as states of doneness for various products. Make sure the tip of the thermometer is not touching bone or fat.

For meat roasts, place the thermometer as near as possible into the center of the roast and in the thickest part. Estimate depth. When roasting poultry, insert a thermometer into the center of the inner thigh muscle. A boneless poultry roast should be treated the same as a meat roast.

To check a steak or other smaller cut, it is best to cut it along the bone or in the center and check the interior color. Rare meat will be somewhat pink. The juice from pork or poultry should never be pink.

The doneness of poultry can be tested by pressing the fleshy part with the fingers. Poultry is done if the meat feels soft.

Beef can be cooked to rare, medium, or well done. Lamb should be cooked to medium or well done. Veal is usually preferred well done. Pork and poultry should always be cooked well done.

Many cases of food illnesses are caused by whole or part carcasses bought from a local small-scale dealer. Improved animal raising and meat-handling regulations and inspections have eliminated most parasites and diseases in meat. This pertains to meat that passes federal inspection and is allowed to travel in interstate trade.

Wild game can also be a lethal source of infection and poisoning. Any wild meat can carry parasites. *No wild meat should ever be cooked and served rare.*

Only excellent quality meat, both domestic and wild, should be used for canning. It should be handled quickly and in total cleanliness, because bacteria can grow at a frightening rate in meats and poultry if given half a chance. Meat picks up bacteria quickly, so it must not be kept at room temperature until it can be handled.

If there is a large amount to be processed, it should be stored temporarily in the refrigerator or in a meat cooler.

And how many times have we read that we should never stuff turkeys or other poultry and store in the refrigerator the night before cooking? If we do, those same *Salmonella* organisms are going to be given the chance to reproduce under extraordinarily good circumstances in the warm dressing inside the poultry. Furthermore, by the time we finally put the poultry into the oven to cook at a moderately low 325° F., it will take the chilled stuffing longer to reach a safely hot temperature, because it has had a chance to get cold.

Any meat that is stuffed presents a need for extra caution. Shortcuts don't pay off . . . except in terms of sickness and misery. Meat, fish, or poultry should be stuffed just before it is to be cooked. The stuffing should be put in lightly, without packing, to allow heat to penetrate more quickly throughout, and thus prevent spoilage.

Stuffing should always reach a temperature of at least 165° F. during roasting. To get a true temperature, a meat thermometer should be placed in the stuffing after the roasting is completed, and should be left there about five minutes. If 165° F. has not been reached, the meat should be cooked longer. Any stuffing cooked separately in the oven should also reach at least 165° F.

If the stuffing is made in advance, it should be stored separately in the refrigerator.

Any leftover stuffing should be carefully removed from the cooked meat, poultry, or fish, and it should be put in a separate container before cooling and storing in the refrigerator. Gravies should be stored separately.

Meat, poultry, or fish should never be partially cooked one day and finished the next.

After poultry is cooked it should never be allowed to stand at room temperature for very long. What does this do to the Thanksgiving and Christmas traditions of serving turkey hot at the main meal, then allowing it to set out while we visit? No wonder there are so many people ill after holidays!

THE LIFE-SAVING RULES

The standard and safe rule of thumb to observe is that cooked dressings and poultry should not be allowed to stand at temperatures between 40° and 120° F.

Of course, poultry is not the only offender. Any meat should be refrigerated as soon as possible after it is served. It can be dangerous if not served and/or stored properly, because bacteria grows rapidly in freshly cooked meat.

As a matter of fact, all leftovers should be refrigerated immediately after a meal.

There aren't many organisms that can survive the brutal experience of a rolling boil. It makes good sense to heat leftover gravy to this point, because gravy and leftovers made with gravy provide a perfect growing medium for disease organisms.

All leftovers should always be served piping hot, and should not be carelessly heated to only a lukewarm temperature. Any frozen or refrigerated leftovers may contain more bacteria than when originally cooked.

Since thorough cooking kills *Salmonella* "germs," the USDA recommends that foods should be cooked above 140° F. at times specified for the particular food (or at higher temperatures for shorter periods).

There are also rules we must know if we are to use the common, ordinary egg safely. Eggs may be laden with salmonellae.

(This information has been available for years, but few of us have known about it.)

For safe storage, it is important to keep fresh eggs clean and cold. Leftover egg yolks or whites should be stored in covered containers, and they should be used in a day or two.

Only clean eggs with sound shells should be used in any recipe in which the eggs are not thoroughly cooked. This would apply to egg-milk drinks such as eggnog, soft-cooked eggs, poached eggs, scrambled eggs, omelets, uncooked salad dressings, ice cream, meringues, soft custards or puddings cooked on top of the range.

Cracked or soiled eggs stand an excellent chance of containing harmful bacteria. However, they don't have to be thrown out. They can be used in foods which are thoroughly cooked, such as oven-

KEEPING FOOD SAFE

baked goods or foods which are cooked a long time on top of the range. The higher temperatures will destroy most bacteria that might be in the cracked eggs.

Generally, dried or frozen eggs should be used only in foods that are to be thoroughly cooked unless the label states that the product is safe to use in uncooked or slightly heated foods.

Pastry mixes such as cake and cookie mixes often contain dried eggs, which may contain dormant salmonellae; therefore, it is important to wash anything these mixes touch such as bowls, spoons, utensils, and hands. It is also very important never to taste an uncooked pastry mix.

Foods containing a large quantity of eggs should be cooled quickly if they are not to be served hot. They should not be allowed to cool off slowly at room temperatures, because this may give bacteria a head start. To speed cooling of custards and puddings, set them in ice water and stir, then refrigerate until serving time.

On picnics, egg-rich pastries should be kept well chilled in portable coolers. This includes cream, custard, or meringue pies and cakes, cream puffs and eclairs with custard fillings. To be on the safe side we should use less "risky" desserts on picnics.

In general, any food which tends to spoil easily without refrigeration should not be exposed to warm temperatures for more than half an hour before it is eaten.

The classic horror stories about illness caused by potato salad at the church picnic points up the fact that anything containing eggs should not stand at temperatures between 40° and 120° F. This rule also applies to sandwich fillings containing meat, fish, eggs, or mayonnaise.

Bacteria can grow alarmingly fast between temperatures of 60° and 120° F. Some bacteria can grow between the temperatures of 40° and 60° F.

Obviously, temperature control is the most effective way to protect food against hazardous microbial growth.

The length of time that food is held at these temperatures is also important, because the time-temperature relationship determines what happens to any microbial contaminants. Storage of frozen food at 0° F. or below prevents most microorganisms from growing and even kills some cells, but it does not destroy all of them.

Thawing causes microbial growth and chemical changes to accelerate as the temperature rises. Unless thawed foods are used promptly, they will begin to spoil and eventually may become too dangerous to eat.

Anytime we leave frozen meat, fish, or poultry out overnight to thaw we run the risk of illness. We run an even greater risk when we leave food out all day to defrost for dinner in the evening.

Of course frozen meat, poultry, or fish may be cooked without thawing, but enough time must be allowed to be certain the center is well cooked and safe to eat. Primal cuts of beef do not have to be well done to be safe, but should be hot. Frozen meats, poultry, and fish take at least one and a half times as long to cook as thawed products of the same size and shape. Frozen combination meat dishes (such as casseroles) should not be thawed before cooking.

The proper place to defrost foods is in the refrigerator or in waterproof plastic containers under running water or in a container of cold water. When defrosting, meat or poultry must not come in contact with other foods. Any blood may contaminate.

Foods that must be refrigerated include perishable food products and potentially hazardous foods. The potentially hazardous foods are those that consist in whole or in part of milk or milk products, eggs, meats, poultry, fish and shellfish. These include cream pies, custards, and salads made of meat, poultry, fish, eggs, and milk products.

Refrigeration at 35° to 45° F. is effective for short-term storage. The danger zone lies between 45° and 115° F., where both infectious bacteria and toxin-producing microorganisms grow rapidly. Food may not be safe to eat if held for more than three or four hours at temperatures between 60° and 120° F. This includes the preparation, storage, and serving time.

Holding perishable foods at temperatures in the danger zone should be avoided or kept to the minimum time necessary for preparation and serving.

KEEPING FOOD SAFE

The basic rule for safe handling of all perishable foods is to keep hot foods HOT (above 140° F.) and cold foods COLD (below 40° F.).

Food should be served soon after cooking, or it should be refrigerated promptly.

Microorganisms can be destroyed by heat. Most molds and yeasts will be killed at temperatures of 140° to 190° F. Most bacteria can also be destroyed at these temperatures.

Holding foods for several hours in an automatic oven prior to serving is not safe if the food is in the temperature zone of 60° F. to 120° F. for more than three or four hours.

Since buffet foods are often held for lengthy periods at room temperatures, they are more susceptible to food poisons or infections. A candle or alcohol burner may not keep food hot enough to control harmful bacteria. The electric hot tray is better . . . if it can maintain temperatures of 140° F. or above.

The directions on the packages of all prepared and partially prepared frozen foods must be followed exactly. Generally speaking, in high altitude areas frozen foods need extra cooking time because the boiling temperature of liquids is lower.

If all of this stressing of the importance of cooking foods properly hasn't impressed us sufficiently to be wary of eating raw meat or fish, now is the time to stop and think of the risks involved. All meat contains some form of bacteria.

Remembering the cautions about consuming dried eggs, we should avoid licking beaters or spoons when mixing up cake or cookie mixes.

And while we're on the subject of tasting, it's a good time to point out that cooking utensils should never be used for tasting food. Fingers should never be licked when working with food, and tasting should always be done with a clean spoon.

Before any home-canned meats or vegetables are tasted, they should be boiled in an open kettle for 20 minutes. Even commercially canned foods should be brought to a boil before eating.

Detecting spoilage in home or commercially canned foods is largely a matter of using our eyes, noses, and our common sense.

THE LIFE-SAVING RULES

However, some dangerous foods give no noticeable signs. Therefore, all suspect food should be boiled for 20 minutes before tasting.

Leakage from jars, patches of mold, and a foamy or murky appearance are visible signs the product should be discarded. The odor from an opened jar or can should be pleasant and characteristic of the product. The food should not be used if it doesn't look right or smell right, or if the ends of the can bulge, or if there is leakage or spurting liquids, off-odor or mold. It should not even be tasted, but destroyed immediately so that not even an animal can consume it.

One of the foods most of us would least suspect of causing foodborne illness would be ham, because ham keeps longer than many other meats. However, ham is the food most often associated with "Staph" illness.

"Staph" organisms themselves cannot hurt us. In fact, they are in the air we breathe and on our skin. But under certain conditions, like in the pudding we made and left at room temperature, "Staph" organisms can multiply rapidly and produce a poisonous toxin. For this reason, ham should be kept hot or cold, never in between.

Ham can basically be divided into two groups: those which must be cooked and those which can be eaten as is.

Hams labeled *only* "cured" or "smoked" must be fully cooked in our own kitchens before serving. The odds are high that illness would result from eating an uncooked ham. Cured or smoked hams must be cooked to an internal temperature of 160° F. Fresh hams are the same as pork, and they must be cooked to an internal temperature of 170° F.

Hams labeled "fully cooked" and canned hams (which are thoroughly cooked in processing) are ready to eat. To be more delicious, they should be heated to an internal temperature of 140° F., but they are safe to eat cold if they have not been contaminated after opening. In any case, ham should not be left at room temperatures for long periods of time.

USDA meat inspectors suggest that any ham not labeled specifically "fully cooked" should be cooked before it is eaten.

The old standby in the meat counter . . . hamburger . . . is potentially one of the most dangerous meats we can bring into our kitchens.

However, there's no need to be nervous. Hamburger is safe if simple rules are observed.

The problem is that hamburger, being ground up, has many more surfaces which are exposed to air and bacteria. In other words, the surfaces where bacteria can grow have been multiplied many times over. Mishandling hamburger gives bacteria the chance to grow.

If hamburger must be thawed quickly, it should be put in a watertight wrapper in cold water or in a closed double paper bag at room temperature. (The double paper bag keeps the surface cool.)

Only clean hands should be used to handle hamburger. Hands and utensils should always be washed with hot water and soap both before and after handling hamburger.

Any cuts or irritations on our hands should be covered with plastic gloves, and it is essential to keep hands away from noses, mouths, hair, and any skin infections when we're working with food.

If there is any delay between the time of preparation and cooking, hamburger should be put back into the refrigerator.

High bacterial counts are not necessarily a hazard to health as long as the meat is cooked thoroughly before it is eaten and proper handling methods are used. Ground beef, often made from trimmings, has been handled more than other cuts of meat. The bacteria in ground beef are not necessarily harmful, but even if they do not cause illness, they may affect the quality and perhaps spoil the meat if it is not handled properly.

Any discussion of the life-saving rules would not be complete without some consideration of water safety. In the event of an interruption of our normal water supply, in an emergency or during a future water shortage, we should know how to purify the water we drink and use.

Water can be made safe by boiling for at least five minutes. An even simpler way to purify water is with halazone tablets. Two halazone pills should make a quart of water safe to drink in half an hour.

Once the most dependable chemical disinfectant for water was thought to be chlorine; however, recent research has indicated possible harmful effects from the use of chlorine in drinking water.

Tincture of iodine can also be used as a water purifier. One drop of fresh iodine will make a quart of water safe to drink in half an hour.

If this knowledge could have been magically given to prehistoric man, how would history have been changed? How far have we come from that day when prehistoric man dropped a piece of raw flesh into the fire? Have we really liberated ourselves from the foodborne diseases which plagued early man?

Even though we may have some of the world's greatest supermarkets and kitchens and an almost unlimited source of miracle drugs and cleansers, one salient factor still remains. No amount of food artistry or skill in cooking can be anything more than a sham if the food is not prepared under sanitary conditions and is free from disease.

TABLE 2

RECOMMENDED INTERNAL TEMPERATURES
FOR MEAT AND POULTRY COOKING*

FRESH BEEF	
Rare	140° F.
Medium	160° F.
Well done	170° F.
FRESH VEAL	170° F.
FRESH LAMB	
Medium	160° F.
Well done	170° F.
FRESH PORK	
Loin	170° F.
Other roasts	185° F.

* Further information may be obtained in Home and Garden Bulletin 1, *Family Fare: Food Management and Recipes*. (Single copies may be obtained by sending a request to: Office of Information, U. S. Dept. of Agriculture, Washington, D.C., 20250.)

CURED PORK (cook before eating)
Ham   160° F.
Shoulder   170° F.
Canadian bacon   160° F.

CANNED PORK (fully cooked)
Ham   140° F.

POULTRY
Chicken   165° to 170° F.
Turkey   180° to 185° F.
Boneless roasts   170° to 175° F.
Stuffing   165° F.

TABLE 3

CRITICAL TEMPERATURES
FOR SAFE FOOD HANDLING

Operation	Internal Temperature ° F.
Home canning	240° F. to 260° F.
Cooking	165° F. or higher
Warm holding	140° F. or higher
DANGER ZONE	45° F. to 115° F.
Refrigeration	35° F. to 45° F.
Frozen storage	0° F. or lower

3

The Spoilers

Food spoilage has always been one of man's most frustrating and infuriating problems.

Since those misty days in prehistory when the first human beings shared an animal carcass on the floor of a cave, the problem has been with us.

There is little doubt that prehistoric human beings experienced the same problems we encounter today from spoiled food.

If a cave dweller was lucky enough to tear off part of the flesh of the animal without his companions' noticing his action, he ate until he couldn't hold any more. Then, more than likely, he hid his leftovers under a rock. At first the meat was tempting to him; soon, however, it turned a darkish color, then green, and finally it smelled so foul the bewildered and frustrated cave dweller either had to throw it as far away as possible, or burn or bury it to rid himself of the horrendous odor.

Now, of course, we know that the molds, yeasts, and bacteria present in the meat, on the dirt floor, and in the stale air of the cave were digesting the meat (or decomposing it). We understand now that the enzymes in the meat were altering its chemical composition.

It may have taken thousands of years for mankind to realize that if he ate spoiled food, his chances of getting very sick and even dying were enormous.

It would seem that anything as disagreeable as spoiled food would be easy to define. Quite the contrary, however. "Spoilage" is difficult to pin down, to define, because people have different concepts about the edibility or fitness of foods for eating.

We usually think about spoilage in terms of decomposition. But this is not at all satisfactory, because it leaves out food which, although not decomposed, harbors certain kinds of organisms or toxins in such great numbers the food is totally unfit for consumption.

Enzymes, or substances which hasten chemical changes in all living things, can also cause undesirable results when uncontrolled.

Generally, food spoilage can be broken down into definite categories:

(1) Undesirable and advanced stages of maturity.
(2) Pollution at some stage during production and handling.
(3) The presence of objectionable chemical and physical changes caused by oxidation, by enzymes, the activity of microbes, insects, rodents, parasites, and damage from pressure, freezing, humidity, moisture, heating, and/or drying.
(4) The presence of microorganisms and/or parasites causing foodborne diseases.

There are certain indications of spoilage which make food unpalatable, but not hazardous to health. These are the rancid odor and off flavor of fats caused by oxidation, the slime on the surface of meat, and the fermentations of fruit juices due to yeast growth. All of these signs of spoilage are repulsive, but they should not make us physically ill.

(Oxidation is a chemical process in which oxygen combines with food constituents and results in the production of undesirable flavor and color. In some cases oxidation occurs without the aid of enzymes.)

But it is an entirely different matter when we encounter the signals which indicate bacterial spoilage. Off odors and sour tastes in bland foods are genuine danger signals. However, by the time a food develops these two dangerous characteristics, it is very unwise for us to experiment by tasting it.

Even more frustrating, of course, is the food which has become lethal, but which has no off odor or unusual taste. Unfortunately, the most deadly of all foodborne diseases, botulism, gives no warning of spoilage.

For this reason, we are forced to use great care in our canning and preserving measures and in cleaning and preparing foods for the table.

It wasn't until almost three hundred years ago that scientists began discovering what actually caused food to spoil.

In Delft, Holland, in 1688, a nondescript janitor named Antony van Leeuwenhoek had become obsessed with grinding lenses in order to enlarge the appearance of the tiny specks he found in the water of his rain barrel.

Eventually, van Leeuwenhoek perfected his lenses to the point where he could identify the specks as living organisms. He was the first human being to actually see the molds, yeasts, and bacteria which live on everything around us, even in our air and the water we drink.

Van Leeuwenhoek had opened the door for the perfection of the microscope.

In the eighteenth century, Lazzaro Spallanzani, an Italian biologist, dared to disagree with the learned men of his day. He proved the tiny "animalcules" made visible by the microscope did not arise spontaneously from decaying matter. He proved once and for all that the microorganisms which inhabit decaying matter have come from parents and in turn will become parents themselves.

He further proved that bacteria can make food spoil and, more importantly, that their spread can be controlled.

Louis Pasteur, of course, is best remembered for his famous studies in fermentation which eventually brought about many changes in brewing, distilling, and wine making, and thus fame to France's great vineyards. He is remembered, too, for his sterilization system which stops the fermentation in foods by killing the microbes in the liquids.

The works of these early scientists have made it possible for today's food manufacturers and processors to avoid spoilage by taking certain precautions.

Unfortunately, this same knowledge is not as available to consumers. If we understand how the microorganisms grow and how they and enzymes work to cause spoilage, we can prevent a great deal of waste in our own homes.

Molds

We are inclined, when confronted with the blue mold on a loaf of French bread, to shudder and hurriedly discard the loaf. The only thought we give to the mold is that it is disagreeable, and we want to be rid of it as soon as possible.

Until recently it has been assumed that common molds were not harmful, and that it was only our squeamishness which dictated the discarding of moldy foods. Moldy food, unfit for humans, often was fed to animals.

But in recent years it has been found that some common molds, under conditions not yet fully understood, can produce a toxin (aflatoxin) that is very harmful to man. Intensive research is being done on toxins produced by molds. Aflatoxin is linked with liver cancer in animals. It has not been firmly established that aflatoxin produces liver cancer in man.

Toxin-producing strains grow on cereal grains, peanuts, and various tree nuts.

We should throw away all moldy foods, except mold-ripened cheeses and other mold-ripened foods where the mold has been deliberately introduced and is known to be safe (9).

It is not enough to trim or sort out portions of food where we can see the mold. The visible mold is only the spore-bearing part of the organism, and most of the mold lies imbedded in the food tissues where its presence is not obvious (11).

But not all molds are villains. One of the greatest "miracle drugs" discovered in modern times is penicillin. Certain cheeses, such as Roquefort, Camembert, and the delightful Brie, owe their flavor to a mold which grows in and ripens them. Many Oriental foods such as soy sauce and miso have been made possible by the growth of certain other molds. Some, grown for food or feed, are used to make products like amylase for bread making or citric acid used in soft drinks.

Once mold growth is under way, its pace can be astounding. The beginning of the growth of molds is slow compared to that of bacteria or yeast; therefore, when conditions are favorable for all of these organisms, molds usually lose out in the competition.

In general, molds require oxygen for growth; this is true at least for the molds growing on foods. The majority also favor an acid medium. Most molds need less moisture for growth than yeasts and bacteria. For this reason, it has been claimed that below 14 to 15 percent total moisture in flour or in some dried fruits will prevent or greatly delay mold growth.

Most molds grow well at ordinary temperatures. They seem to grow best around 77° to 86° F., but some grow well at 95° to 98.6° F., and some do well at still higher temperatures. Others grow fairly well at freezing temperatures, and some can even grow slowly at temperatures below freezing. Growth has been reported as low as 23° to 14° F.

We are all familiar with mold's fuzzy or cottony appearance. Molds are tiny, simple plants which belong to the fungi group. They are closely related to the mildews, rusts, and mushrooms. Molds have no chlorophyll and, therefore, they cannot manufacture their own food. They must live on food made by other plants, animals, or on decaying matter. Some molds live as parasites on insects and even on certain fungi.

If we didn't think of molds in disagreeable terms most of the time, the colors would be very beautiful to us. Some of the characteristic hues are red, purple, yellow, brown, gray, black, green, blue-green, orange, pink, and lavender.

Some molds are loose and fluffy; others are compact. Some look velvety on the upper surfaces, and some are dry and powdery. Others are wet and gelatinous. Some molds are restricted in size, while others seem to be limited only by the food or the size of the container.

Black mold, the group also known as bread mold, gets its name because all of these molds produce dark-colored spores. Common bread mold forms a cottony, soft white growth on damp bread. A group of molds known as the blue molds also grow on bread.

A green mold grows on various kinds of cheeses and can spoil them. Other molds, called water molds, are often found in water and soil.

Molds develop from tiny particles called the spores. When the spore settles on a damp food substance such as bread, it swells and begins to grow by producing many tiny threads. Some of the threads are like roots, while others spread out and branch many times.

As the plant body of the mold matures, many of the upright threads produce spore cases which contain thousands of spores. Each spore case is called a sporangium and is about the size of a pin head. When the spore case matures and breaks open, the spores are set free and are carried away by air currents. These spores in the air settle on damp foods and develop into new molds.

MOLDS COMMONLY FOUND IN FOODS

Rhizopus, the so-called bread mold, is very common and it is involved in the spoilage of many foods: berries, vegetables, bread, etc. This type of mold is characterized by an abundant, black cottony mass of branching and intertwined filaments.

Asperquillus, niger group, is frequently found in foods. The spore-bearing heads are large, tightly packed, and globular, and they may be black, brownish black, or purple-brown.

Penicillium is widespread and also important in foods. The genus is divided into groups and subgroups, and there are many species. *Penicillium camemberti* and *P. roqueforti* with blue-green tints are used in the ripening of blue and Camembert cheeses.

Trichothecium is a pink mold which grows on paper, fruits, and vegetables such as cucumbers, cantaloupes, apples, and peaches.

Geotrichum or "dairy mold" produces a white, cottony growth on dairy products and tomatoes.

Sporotrichum is found growing on chilled meats, where it causes "white spots."

Alternaria are common causes of the spoilage of many acid foods, especially fruits and tomatoes. They form a mass of dirty-looking, gray-green mold.

Molds are easily killed by moist heat. A temperature of 160° to 180° F. for 60 minutes will usually kill most mold spores.

Experimental results have shown that a temperature of 212° F. for several minutes will destroy a large percentage of spores. As a rule, molds are not involved in canned food spoilage, because of the unfavorable conditions for their growth in an airtight container.

Moreover, mold spores are unable to survive the temperature used in the processing of most foods.

Yeasts

Yeasts are microscopic plants which cause spoilage by fermentation. Fermentation, as many of us agree, is great in wine and beer but something else again in applesauce. (During fermentation yeasts help change sugar to alcohol.)

The yeast with which people are most familiar is a substance bakers put into dough to make it rise. This yeast contains a mass of tiny, one-celled plants.

Yeasts increase very rapidly, and the tiny plants are floating in the air almost everywhere. These little organisms which have plagued and pleased us through the centuries are abundant in nature. They are common in orchards, vineyards, and in the soil. Besides existing beautifully in the very air we breathe, they get along well in the intestines of animals and certain insects.

They are most active in foods which contain sugar, and they are also responsible for the bubbles which rise with an alcoholic aroma when we open spoiled cans of fruits. Usually, the seal of a can or jar has been broken by the carbon dioxide gas the yeasts produce.

Yeasts are undesirable when they cause spoilage of sauerkraut, fruit juices and sauces, syrups, molasses, honey, jellies, meats, and other foods.

Like molds, yeasts have no chlorophyll, and they are dependent on other plants and animals for their energy.

Yeasts require less moisture for growth than do most bacteria, but they need more than molds do. The range of temperature for growth of most yeasts is similar to that of molds, with the ideal temperature being around 77° to 86° F. The maximum temperature they can tolerate is 98° to 116° F.

Yeasts are the main product in the production of wines, industrial alcohol, and other products. Yeasts also aid in the production of flavors, of "bouquet" in wines, and leavening of breads.

YEASTS COMMONLY FOUND IN FOODS

Saccharomyces are used in many food industries for the leavening of bread, as top yeasts for ale, for wines, and for the production of alcohol, glycerol, and invertase.

Zygosaccharomyces are notable for their ability to grow in high concentrations of sugar, and they are involved in the spoilage of honey, syrups, and molasses and in the fermentation of soy sauce and some wines.

Debaryomyces are round or oval yeasts which form pellicles on meats.

D. kloeckeri are commonly found on cheese and sausage.

Hanseniaspora are lemon-shaped yeasts which grow in fruit juices.

Candida usually form films and can spoil foods high in acid and salt.

C. lipolytica can spoil butter and oleomargarine.

Brettanomyces produce high amounts of acid and are involved in the fermentation of Belgian lambic beer and English beers.

Rhodotorula are highly visible, red, pink, or yellow yeasts which may cause discoloration on foods (such as the colored spots on meats or pink areas in sauerkraut).

Bacteria

Bacteria, which Leeuwenhoek found so astonishing and intriguing, are all around us in our environments. Some produce spores which can also be carried by the air.

Bacteria are actually single-celled protists made up of living matter called protoplasm. Bacteria, like molds and yeasts, usually do not contain chlorophyll. They are so small that a single round one of a common type is about 1/25,000 of an inch across. When magnified 1,000 times it looks no larger than a pencil point.

Bacteria are found in either an active or a resting form. In the active or vegetative stage, bacteria in food are destroyed at boiling temperature, but those that produce spores go into a resting or dormant state and have been known to live after being subjected to boiling water for several hours. They can survive freezing temperatures and extensive periods of dryness, only to spring back to life, sometimes weeks or months later, when favorable conditions have been restored.

Bacteria grow and reproduce rapidly under good conditions by dividing into two. These two divide into four, four into eight, eight into sixteen, and so on. A *single* bacterium can grow and divide into two new cells in less than half an hour. When division (or fission) takes place every hour, one bacterium can produce over 18,000,000 new bacteria in 24 hours. At the end of 48 hours there will be hundreds of billions of bacteria. In numbers like these, bacteria can cause extensive changes in foods.

There must be enough warmth, considerable moisture and food for bacteria to grow well. Certain kinds of bacteria need oxygen to grow, while others grow best where there is no oxygen.

Some bacteria cause fermentation. (Milk turns sour when certain bacteria are allowed to grow in it.)

Types of bacteria vary with different foods. The time of year, the locality, and the conditions under which the food is produced also determine the types of bacteria we may encounter. Some of the most heat-resistant bacteria are in the soil, so we need to take special care when we prepare, process, serve, and store such foods as spinach and snap beans.

Each kind of bacterium has a definite range of food requirements. For some species that range is wide, but not for others.

These differences and confusing similarities make dealing with bacteria a seemingly complicated task. But, in general, there are some conditions which apply to them all.

Most bacteria grow best in a nearly neutral medium, although a few are favored by an acid reaction. (A few can grow in either an acid or an alkaline medium.)

Natural foods may contain compounds that inhibit some organisms more than others, for instance, the benzoic acid in cranberries.

Certain edible substances slow down or stop the growth of bacteria, so they are frequently added during the processing of foods; for instance, the propionates added to bread to inhibit the growth of molds and rope bacteria.

BACTERIA IMPORTANT IN FOODS

Acetobacter: Family Pseudomonadaceae; 7 species. It is commonly found in grain mash, mother of vinegar, beer, wines, and souring fruits and vegetables.

Achromobacter: Family Achromobacteraceae; 15 species. Many ferment glucose and other sugars, but produce no gas. They are significant in the spoilage of meats, poultry, and seafoods.

Aerobacter: Family Enterobacteriaceae; 2 species. Both are usually found on plants, grain, in water, and in the intestinal tract.

Alcaligenes: Family Achromobacteraceae; 6 species. They do not ferment sugars, but they do produce alkaline reactions which are also produced in litmus milk. (Litmus milk is a laboratory preparation which is used to detect alkalinity or acidity.) These bacteria are found in all decomposing matter, in unpasteurized milk, and in the intestinal tract.

Bacillus: Family Bacillaceae; 25 species. The dreaded *B. anthracis,* which causes anthrax, is one pathogen included in the family. (Pathogens are disease-causing bacteria and viruses.) The family also includes nonpathogens and they are extremely common in the air, dust, soil, water, and on utensils and in various foods. These bacteria are significant in the spoilage of many foods held at room temperature.

Clostridium: Family Bacillaceae; 93 species. The etiologic agents (causing agents) of tetanus, gas gangrene, perfringens food poisoning, and botulism are members of this genus. They are widely distributed in soils, water, and the intestinal tracts of man and animals. They can also be found in many foods where they may not grow.

Corynebacterium: Family Corynebacteriaceae; 33 species. This group contains the bacteria which causes diphtheria and many other diseases. These organisms exist on wheat, beans, and tomatoes. They are also found in the intestinal tracts of man and animals, and they have been found in spoiling foods of various types.

Erwinia: Family Enterobacteriaceae; 17 species. They are generally found among and on plants where they cause specific areas of vegetation to die. These are the most significant bacteria causing fruit and vegetable diseases in the market.

Escherichia: Family Enterobacteriaceae; 4 species. Their favorite haunt is the intestinal track of man and animals, but they can also be found in many other places in nature as well. When they are found in large numbers in foods, fecal contamination is usually indicated.

Flavobacterium: Family Achromobacteraceae; 26 species. They are found in soils, in water, on fish, and on plants.

Lactobacillus: Family Lactobacillaceae; 3 species. They are usually found on plants and in dairy products. Some are used to ferment milks, and others are used in making cheeses. Still others are used in the testing of B vitamins and amino acids. They are also found on cured and processed meats.

Micrococcus: Family Micrococcaceae; 16 species. They are widely found in nature, on the skin of man, and on the hides of animals, as well as in dust, soil, water, and in many foods. Several species are associated with dairy products, through which they enter processed meats like frankfurters.

Pediococcus: Family Lactobacillaceae; 3 species. They are found on plants, and in milk and dairy products. Some can cause problems in sugar refineries where they form a slime in the pipes. Still others are used in yogurt and sour cream starters, while others are commonly found in cured meat products.

Proteus: Family Enterobacteriaceae; 5 species. These are found in the intestinal tracts of man and animals and on decaying materials in general. They have been found in spoiled eggs and meats, especially those allowed to spoil at above refrigerator temperatures.

Pseudomonas: Family Pseudomonadaceae; 149 species. These are found in soils, water, on plants, and in the intestinal tracts of man and animals. These are unquestionably the most important bacteria in the low-temperature spoilage of foods such as meats, poultry, eggs, and seafoods. Some strains produce fluorescent pigments.

Salmonella: Family Enterobacteriaceae; more than 2,000 serotypes. These bacteria are widely found in nature. The causative agents of typhoid and paratyphoid fevers belong to this group as well as those which cause foodborne salmonellosis in man and many animals. All species and strains of this genus are undesirable in foods.

Shigella: Family Enterobacteriaceae; 8 species. These bacteria occur in polluted waters and in the intestinal canal of man where they cause bacillary dysentery and other intestinal disorders. Their primary means of reaching foods are through polluted water and human carriers. These organisms are undesirable in foods.

Staphylococcus: Family Micrococcaceae; 2 species. Both species are common in the nasal cavities of man and some animals as well as on the skin, hide, and other parts of the body. S. *aureus* produces boils, carbuncles, and an important food-poisoning syndrome in man. Their presence in foods in large numbers is undesirable.

Streptococcus: Family Lactobacillaceae; 19 species. Some are associated with the upper respiratory tract of man. They cause diseases like scarlet fever, septic sore throat, etc. Others are found in the intestinal tract of man and animals, and they tend to be rather common on plants and in dairy products. Some cause mastitis in cattle, while others are important in dairy sour milk, and in the manufacture of cheese. Some produce a food-poisoning syndrome in man. The presence of some species in foods in large numbers may indicate fecal contamination.

Streptomyces: Family Streptomycetaceae; 150 species. These bacteria find their way into many foods such as vegetables from the soil, where they commonly exist. They are also found in the mouth or oral cavity of man. They are active in the destruction of both animal and plant matter at temperatures above the refrigerator range.

Enzymes

Without enzymes there is no life. Yet they are probably the least understood of all the spoilers.

An enzyme is an organic catalyst. This means that enzymes speed up chemical reactions without actually taking part in them. There can be no life without enzymes, because all living things, plants and animals, from the smallest to the largest, depend on chemical reactions for life.

These catalysts are present in all foods and are responsible for the ripening of fruits and vegetables and for making meats tender as they "ripen."

This same useful function, when it is extended and allowed to continue, causes foods to spoil.

That is why we must either destroy or inactivate enzymes when we preserve foods. Heat destroys enzymes, and freezing temperatures inactivate them. Any attempt to preserve food is really an effort to create an unfavorable environment for enzymes, whether these enzymes are natural to the foods or produced by microorganisms.

The effects of the enzymes' work can be seen very quickly in the darkening of apples, pears, peaches, or potatoes, if they are pared and left exposed to the air.

If the air has not been removed from a jar in which peaches or pears are canned, they will become discolored.

Another sign of enzyme activity is an unusually soft texture.

Enzymes play an important role in the food industry. For the baker, the brewer, and the confectioner, they liquefy and change starches into fermentable sugar; they convert sugars and modify proteins.

In fruit juices and wine making, enzymes not only speed filtration, but also increase the yield, and improve and clarify the juice.

KEEPING FOOD SAFE

Each enzyme has a particular job to do and usually cannot do another.

Living cells produce these mysterious enzymes. Even microorganisms produce them. Because enzymes are part of the cells or intercellular tissue, they continue to react with these tissues after fruits and vegetables are harvested or after farm animals and game are slaughtered. Enzymes can be extracted and purified from certain biological materials, such as plant sap and glandular tissues, to provide industrial enzymes.

An important point to remember is that enzymes are part of the food we eat. They are necessary for carrying on complicated life processes.

After the living tissue or food material dies, many of the enzymes continue to function. It is in this stage the food substances may undergo the deterioration and disintegration we usually associate with food spoilage.

Fruit that is removed from the vine or tree before it is mature continues to carry on internal biochemical changes. Enzymes take an active part in this process. Green bananas and tomatoes, for instance, will continue to ripen after being pulled from the plant. Fruit is still alive when it is stored, and it will continue to "breathe" or respire by taking in oxygen and giving off carbon dioxide. This process will continue to go on until the fruit is fully ripe.

This is why precooling of the fruit is important in order to slow down respiration before shipping or storage.

This life process continues in meat long after an animal is dead. We know this is so, because there is a distinct favorable alteration of flavor and tenderness of meat if it is aged properly. This increased tenderness and enhanced flavor of aged meat is actually a controlled enzyme process, and it resembles protein digestion. Of course, no one needs to tell us that if it is allowed to continue, a true spoilage will result.

Microorganisms, like all living cells, owe their activity to enzymes. Certain molds, for instance, liberate lipase, an enzyme, which plays an important role in the ripening of Roquefort and other cheeses.

Bacteria also possess enzyme systems which produce profound

changes in foods. Some are desirable; others are distinctly undesirable.

Clear beer is an enzyme product. Ordinary beer will become cloudy when it cools. When treated with a protein-splitting enzyme, the beer will remain clear, even though it is cool.

Other enzymes are used to improve the strength of jellies by reducing the sugar content.

Today the commercial enzyme industry is expanding into a large-scale operation. However, enzymes were used industrially long before anything was known about their chemical composition.

We only have to think how important wine, cheese, and beer have been to mankind through the centuries, and we begin to understand how much our ancestors depended on enzyme action.

Sources of Contamination

GREEN PLANTS AND FRUIT

We know now that the microorganisms Leeuwenhoek found so interesting are living and thriving all around us. There are bacteria everywhere on the natural surfaces of plants and fruits.

For instance, the surface of an unwashed tomato usually has several thousand microorganisms per square centimeter, while a well-washed tomato may have only 400 to 700.

And the outer tissue of an unwashed cabbage can easily hold 1 to 2,000,000 microorganisms per gram, but washed and trimmed cabbage, only 200,000 to 500,000.

The reason for these astounding figures is that the exposed surfaces of plants are contaminated continually by soil, water, sewage, air, and animals.

Some fruits have even been found to contain some microorganisms in their interior. Organisms have been found in healthy root and tuber vegetables.

ANIMALS

The intestinal tracts, hides, hoofs, and hair of animals usually contain not only large numbers of organisms from soil, manure, feed, and water, but also many important kinds of spoilage organisms.

It isn't at all surprising that feathers and feet of poultry carry heavy contamination from similar sources. Sometimes pathogenic organisms capable of causing human disease, such as *Salmonella,* come from animals.

SEWAGE

When untreated domestic sewage is used to fertilize plant crops, there is a likelihood that raw plant food may be contaminated with human pathogens, especially with those causing gastrointestinal diseases. (The use of such sewage is common in undeveloped countries where it is referred to as night soil.)

Natural waters contaminated and polluted with sewage pass the microorganisms on to shellfish, fish, and other fresh water and sea foods. Treated sewage going onto soil or into water also contains microorganisms, although it has smaller numbers and fewer pathogens than raw sewage.

SOIL

The soil contains the greatest variety of microorganisms of any source of contamination. Soil dust whipped up by air currents and soil particles are carried into or onto foods.

Nearly every important microorganism can come from soil. Modern methods of food handling usually involve the washing of the surfaces of foods, so much of the soil is removed from natural, outer surfaces.

WATER

Natural waters contain not only their natural flora, but also microorganisms from soil and perhaps even from animals or sewage. Surface waters in streams and pools and stored waters in lakes and large ponds vary considerably in their microbial content, from many thousand per milliliter after a rainstorm, to the comparatively low numbers that result from self-purification of quiet lakes and ponds or of running water.

THE SPOILERS

Ground waters from springs or wells have passed through layers of rock and soil to a definite level, and so most of the bacteria, as well as the greater part of other suspended material, have been removed.

Bacterial numbers in these waters can range from a few to several hundred bacteria per milliliter.

Contamination can come from water used as an ingredient in a recipe or from washing foods. It can come from the water used for cooling heated foods and from manufactured ice used for preserving foods.

Usually each food product contains certain microorganisms.

For instance, the gas-forming coliform bacteria may enter milk from cooling tank water and cause trouble in cheese made from the milk. Bacteria causing ropiness and slime of milk come from water and/or soil.

The iron bacteria, whose sheaths contain ferric hydroxide, may gum up an entire water supply and they are very difficult to remove.

AIR

Disease organisms, especially those causing respiratory infections, can easily be spread from human to human through the air. Food products can also become contaminated by the microorganisms floating in the air.

Total numbers of microorganisms in a food product can be increased from the air, especially if that air is being used in the making of the food.

Spoilage organisms come from the air, and so do those which interfere with food fermentations.

Mold spores from the air can cause trouble in cheese, meat, sweetened condensed milk, and sliced bread and bacon.

Air doesn't contain natural microorganisms; all that are present in the air have come there by accident and usually are on suspended solid particles or in moisture droplets.

Microorganisms get into air through dust or lint; dry soil; spray from streams, lakes, or oceans; from droplets of moisture from coughing, sneezing, or talking; from sporulating molds on walls, ceilings, floors, foods, and ingredients; and from sprays or dusts from food products or ingredients.

Microorganisms in the air have no opportunity for growth; they just persist there.

Mold spores and yeasts are also usually present in the air. Bacterial spores are relatively uncommon in dust-free air. It is only common sense to realize that whenever dusts or sprays of various materials are carried up into the air, the microorganisms which are characteristic of the suspended materials will also be present. For instance, soil organisms come to the air from soil and dust, water organisms come from water spray, and plant organisms from feed or fodder dust.

The numbers of microorganisms in the air at any given time depend upon the amount of movement in the environment, the amount of sunshine, humidity, location, and amount of suspended dust or spray. The numbers of bacteria can vary from less than one per cubic foot on a mountain top to thousands in dusty or smoggy air.

Individual microorganisms and those on suspended dust or in droplets settle out in quiet air; moving air brings more organisms up into it. Therefore, numbers of microorganisms in the air are increased by air currents caused by movements of animals, human beings, vehicles, ventilation, and breezes.

Direct rays from the sun will kill microorganisms suspended in the air, and reduce their numbers. Dry air usually contains more organisms than similar air in a moist condition. Rain or snow removes organisms from the air so that a hard, steady rainfall will practically clear the air.

As we think back over the various types of spoilers, we begin to realize that detecting them in our food is largely a matter of using our eyes, noses, and common sense.

Leakage from swollen jars or cans, patches of mold, and a foamy or murky appearance are visible signs the product should be discarded. The odor from an opened jar should be pleasant and characteristic of the food.

THE SPOILERS

In general, we should remember to throw food away if:

(1) It doesn't look or smell right.
(2) The ends of a can bulge.
(3) There is leakage or spurting liquids.
(4) There is a rotted or objectionable odor.

Furthermore, the food should not even be tasted, but destroyed so that not even an animal can consume it.

Now we understand it was the spoilers which ruined the meat our cave friend hid under the rock. We realize the molds, yeasts, bacteria, and enzymes which are so abundant in our environment were responsible for the spoilage of his meat, just as they are responsible for the spoilage of food in our supermarkets and in our homes.

We live side by side with the spoilers. They should command our respect, because they are integral parts of life's processes and they can undermine our health and our food supply.

4

Problem Foods and Safeguards

Once there was a common body of knowledge and superstition which dealt with spoilage and with the care of foods. It was passed down by word of mouth from one generation to the next. But our modern technology . . . our refrigerators and our freezers . . . our miracle drugs and our cleansers . . . have disrupted this flow of passed-down information. Even the superstition is no longer commonly known, much less applied.

It is time we learned to think of the foods in our kitchens on the basis of their stability as nonperishable, semiperishable, and perishable products.

Sugar, that delicacy which has tickled mankind's palate since it was first savored in the mists of history in India, is a perfect example of a true nonperishable food. Very few foods can actually be classified as nonperishables.

When we glance at the colorful rows of canned goods on supermarket shelves, we are likely to think of them as nonperishables. But even canned goods can become perishable under certain circumstances.

The semiperishables are the dry foods, such as flour, dry beans, bakery products, hard cheeses, dried fruits and vegetables. Frozen foods, though basically perishable, are actually semiperishable, if they are freezer-stored properly.

KEEPING FOOD SAFE

The majority of our foods must be classified as perishables. This group includes meats, poultry, fish, milk, eggs, many fruits and vegetables, and all cooked or "made" food items, except the dry and very acid ones.

General guidelines and safeguards can easily be translated into dollars and cents in our food budgets.

As a rule of thumb we should remember that moist foods should be kept moist; dried foods should be kept dry; frozen foods frozen and fresh foods as nearly in their fresh-cut or field-ripened condition as possible.

As we have learned, application of common sense measures as well as technical information on how to avoid spoilage in our foods adds up to better health and more productive lives.

The control of spoilage and foodborne diseases really involves two primary preventative measures in food storage and handling. Those measures are:

(1) The prevention of contamination of the foods.
(2) Conditions must be established in which bacteria, yeasts, and molds cannot grow and under which enzymes cannot produce chemical changes rapidly.

Temperature is the most important single factor governing food stability. It affects the quality of products in every respect, including color, texture, aroma, flavor, and nutritional content. Low temperatures retard quality losses and delay spoilage by slowing the action of enzymes and the growth of organisms. Storage conditions which are excellent for one group of foods may be far from ideal for others.

Generally speaking, fresh, perishable food should be used as soon after harvest or purchase as possible. If storage is necessary, the food should be cared for under the proper temperature and humidity conditions.

We must make every effort to use fresh foods before they undergo any loss of quality. Even under the best of storage conditions freshness and nutritive value can be lost if foods are stored for too long.

Our task of storing perishable foods properly is minimized con-

PROBLEM FOODS AND SAFEGUARDS

siderably if we buy only the amount we can use quickly. We should shop with the intent of using perishable foods immediately, rather than storing them for longer periods.

Many products are now dated to give us an idea how long an item should be sold as well as to let us know when to use it for maximum quality. More and more foods are being dated all the time.

Improved methods of detecting bacterial presence in man and animal are more efficient now than ever before, so federal officials have become acutely aware how often foodborne illnesses may be occurring.

Our eating habits today often depend on bulk preparation and mass distribution of human and animal foods on both the national and international levels. More people are eating more food from fewer kitchens all the time. This factor helps spread contamination very effectively and rapidly.

We are also eating out more often than we used to, so we are exposed more often to the possibility of poor food preparation in quantity cooking.

Meanwhile at home, premixed, ready-cooked, improperly processed convenience foods can be significant sources of food illnesses.

PART 1

Breads, Cereals, and Grain Products

That sack of flour we brought home from the supermarket doesn't look like a landmark of history. Yet, it represents more than four thousand years of progress for mankind.

One of the most significant moments in all of history occurred when the first caveman or woman discovered that seeds of wild wheat were good to eat. Eventually, whole villages sprang up around the edges of wheat fields, and structured society was given a boost.

The wheat that early man came to depend on so heavily most certainly was loaded with dirt and insects.

Even with today's sophisticated harvesting methods, wheat contains stones, chaff, dirt, insects, and animal residue.

Freshly harvested grain contains millions of microorganisms, many of which occur naturally on the grain. Scouring, washing, and milling make the flour as civilized as we like it. Even so, the numbers of bacteria in the wheat flour that we bring into our kitchens can vary from a few hundred per gram to millions. Most samples of wheat flour in our supermarkets contain anywhere from a few hundred to a few thousand bacteria per gram.

A significant spoilage agent in bread is a disease called rope. (Rope spores can survive the baking process.) Rope can also occur in doughnuts, cakes, and brown breads. It causes sourness and stickiness after a few days. There is also an off odor. The bacteria can come from potatoes, flour, yeast, spices, fruit, sugar, molasses, dust, dirt . . . any number of places.

Molds are the most common (and therefore the most important) cause of spoilage in bread and bakery products. Temperatures reached during baking are usually high enough to kill all mold spores, but the bread is recontaminated after baking. Molds can easily reach the bread while it is cooling, especially from the air. They are often present on the wrappers we package the bread in, and they usually begin growing in the creases of the loaf and between the slices.

Another type of spoilage occurring in bread is too much acid fermentation, which is caused by lactic acid and coliform bacteria that are normal in flour pastes or dough.

The fermentation can become too powerful if we allow too much time to pass during the rising period or if we knead the dough too long. (However, in our home kitchens the chances of underkneading are greater than overkneading.)

Bread turns sour if fermentation is allowed to go on too long. Excessive growth of bacteria during the rising or kneading periods may easily destroy some of the gas-holding capacity so essential during expansion of the dough. If this happens, we wind up with a sticky dough, and our finished product will be heavy and not tempting.

One of the most spectacular types of bread spoilage is the red

PROBLEM FOODS AND SAFEGUARDS

or "bloody" spoilage. We can imagine how ancient man could consider the sudden appearance of "blood" as a sign of a miracle. Today, scientists know the red color is caused by the growth of pigmented bacteria. However, this type of spoilage is very rare.

Another uncommon type of bread spoilage is caused by a yeast-like fungus that produces chalk-like spots, or "chalky bread".

SAFEGUARDS

Bread should be kept in its original wrapper and stored in a cool, dry, and well-ventilated container unless the weather is hot and humid. Then it should be kept in the refrigerator to guard against molding. We should wash out our bread container with a soda solution once a week.

We are far from helpless when it comes to fighting mold. There are three excellent ways we can fight mold in bread:

(1) After baking, bread should be cooled promptly and adequately before we wrap it. This reduces the condensation of moisture inside the wrapper.
(2) Bread should not be allowed to set out longer than necessary.
(3) Bread should be kept cool to ward off mold growth. Better yet, it should be frozen until we are ready to use it.

In commercial bread, sodium and calcium propionates (0.1 to 0.32 percent of the weight of the flour used) are often used to inhibit mold growth and ropiness. Calcium propionate is the calcium salt of propionic acid, and it is commonly found in both natural and processed foods.

Calcium propionate will not completely prevent mold, but it will inhibit its growth for about a week (15).

If we are unfortunate enough to get rope disease in our kitchens, there are a number of ways we can deal with the problem. However, once the disease organisms start growing in a kitchen, they are extremely difficult to eliminate. (If we don't make our own bread, rope is not a significant problem.)

To rid a kitchen of rope disease:

(1) Discard any bread that shows signs of ropiness.
(2) Be sure that no bits of dough are left in the mixing bowl or on the bread board.
(3) Scrub every utensil used in making bread, including the bread board, bread box, or other storage places with hot water and vinegar. (Use one tablespoon vinegar to a quart of boiling water.)
(4) Use buttermilk for ¼ to ½ the total liquid used in the recipe. Do not add soda.
(5) In place of the buttermilk, use a small amount of white vinegar. Add to the liquid in the bread recipe. Use no more than 1 teaspoon vinegar to each cup of liquid.
(6) Freshly baked loaves should be cooled promptly after baking.
(7) Bread should be stored at a cool or freezing temperature. (Freezing bread and storing it in the frozen condition will prevent ropiness entirely.)

But let's assume for the moment that our problems with bread are no more serious than figuring out how to keep hard rolls hard and soft rolls soft. (This can be an exasperating problem.)

Fresh, hard or crisp-crusted breads or rolls should be frozen in loosely wrapped freezer materials, or we can package them in a heavy plastic bag. When we want to use them we can unwrap the rolls and thaw at room temperature, then the bread should be heated (uncovered) in a hot oven for five minutes and served immediately.

Frozen, soft-crusted bread and rolls can be wrapped in aluminum foil and heated for a few minutes in a moderate, preheated oven. (Freezing stale or partially stale bread will not restore freshness.) When thawing them at room temperature, soft-crusted breads should be left in their wrappers.

All breads, cereals, and grain products should be stored in containers which will keep out insects and rodents and which will help protect against rapid changes in temperature and moisture. (Suddenly, the old-fashioned tin bread box begins to look better and better.)

PROBLEM FOODS AND SAFEGUARDS

Those cereals the children crave should be kept tightly covered to keep them from either drying out too rapidly or absorbing too much moisture. Opened packages need to be resealed tightly or the contents should be transferred to tightly covered containers.

Summer is not the time of year to buy grain in large quantities, because insect infestation is much more likely in warm weather.

PART 2

Sugars and Sugar Products

When we think of sugar, most of us think only of beet or cane sugar, which is sucrose. But there are also molasses, syrups, maple sap, maple sugar, honey, and candy.

CONTAMINATION

Sucrose, the raw juice expressed from sugar cane and beets, can become very high in microbial content unless processing is prompt. Most of the microorganisms are those from the plant itself and from the soil. If organisms grow to any real extent, inversion of sucrose or even destruction of the sugar can take place. And although certain stages of refining will kill yeasts and vegetative cells of bacteria, bacterial spores remain alive.

Even more microorganisms may contaminate the sugar during the actual bagging process before the sugar is sent to market. In raw sugars 400 to 68,000 organisms per gram have been found, and in molasses from 100 to 190,000 per gram, although numbers vary considerably.

Granulated sugars now on the market are very low in microbial content; most contain from a few to several hundred organisms per gram, and these are mostly bacterial spores.

Maple sap in the tree is practically sterile, but it becomes contaminated from outside sources through the tap holes and by the spout which is inserted into the trunk of the tree. But the use of plastic tubing and plastic bags which let bacteria-killing sunlight into the syrup will reduce the microbial content greatly.

The chief sources of microorganisms in honey are the nectar in flower and yeasts, and (this boggles the mind) bacteria from the intestinal tracts of bees.

Candies from retail markets have been found to contain from 0 to 2,000,000 bacteria per piece, but most pieces have no more than a few hundred. Candies are contaminated more from their ingredients than from the air, dust, or handling.

SPOILAGE

The spoilage of sugars or concentrated solutions of sugars is limited to those microorganisms with a real sweet tooth which are able to grow in such high concentrations. These are species of *Pediococcus* and *Bacillus,* certain yeasts and some molds. As the sugar concentrations decrease, increasing numbers and kinds of organisms can grow.

Raw cane or beet juice is not high in sugar and contains a good supply of food for microorganisms, and therefore is easily spoiled by numerous organisms if enough time is allowed to pass. Until the clarification process, gum and slime may form which can cause pipes, strainers, and pumps to clog. Sucrose can be completely destroyed by inversion, fermentation from acid, or oxidation by bacteria.

Microbial spoilage of molasses is not common. However, certain yeasts may survive the heating process and gaseous chemical spoilage can take place. Molasses or syrup exposed to air will eventually mold on the surface.

Sap from the sugar maple becomes contaminated when it is drawn. Although a moderate amount of microorganism growth may improve flavor and color, the sap often stands under conditions that favor excessive growth and then spoilage results. There are four chief types of spoilage.

Honey is variable in composition, but must contain no more than 25 percent moisture. Because of honey's high sugar content (70 to 80 percent) and its acidity (pH 3.2 to 4.2), the chief cause of its spoilage is osmophilic yeasts. Most molds don't grow well on honey, although species of *Penicillium* and *Mucor* have been found to develop slowly. The fermentation process is usually slow, lasting for months, and the chief products are carbon dioxide, alcohol, and nonvolatile acids which give an off-flavor to the honey. Darkening and crystallization usually occur along with the fermentation.

Most candies are not subject to microbial spoilage because of their comparatively high sugar and low moisture content. Exceptions, of course, are the chocolate with soft centers of fondant or inverted sugar, which can burst or explode. Yeasts or species of *Clostridium* growing in these candies develop a gas pressure which may push out some of the syrup or fondant through a weak spot in the chocolate covering. Molds have also been found growing under the coating of chocolate bars.

SAFEGUARDS

Sugar is not at all difficult to store and keep, as the ancient Arabs well knew when they used it almost like money for trading in Spain centuries ago. And of course, we know now the Arabs' sugar kept well because sugars of most types generally have such a low moisture content that microorganisms simply cannot grow. It is only when moisture has been absorbed that there is any real chance for microbial spoilage.

A safe rule of thumb for us to remember is that sugar will keep well if we keep it as dry as possible. Of course, we must take common sense precautions like making sure storage conditions are good enough to keep out sugar-craving insects like the ant, as well as rats and mice. Storage temperatures of 40° to 50° F. are best.

And because of their high sugar concentrations, most candies will not be spoiled by microbial growth. But the soft fillings of chocolate-covered candies are moist enough to furnish a sweet hideaway for microorganisms. The heavy coating of chocolate we love on cream-filled candies serves the very useful function of preventing additional microbial contamination and helps keep the candy from bursting. Fondants or other fillings are selected so they will not permit the growth of gas-formers.

The boiling process during evaporation of maple sap to maple syrup kills the most important spoilage organisms.

In order to keep our syrups and molasses for longer periods we should periodically turn the bottle upside down a few times to allow the product to mix well. Pouring the contents of half-empty bottles into smaller containers also helps prolong the usable period of syrups and molasses.

Honey distributed on a fairly small scale is usually not pasteurized. It may crystallize and spoil from osmophilic yeasts. Commercially distributed honey is usually pasteurized at 160° to 170° F. for a few minutes.

This means we can treat honey in our own kitchens by setting the container in a pan of water and heating it rapidly to at least 160° F., and by holding it there for five minutes and then promptly cooling it to 90° or 100° F.

Jellies and syrups should be stored at room temperature until they are opened. After that, honey, syrups, and jellies are better protected from mold in the refrigerator.

PART 3

Fruits and Vegetables

CONTAMINATION

When we look at a magnificent apple tree covered with juicy red fruit, the last thing that crosses our minds is bacterial contamination. But just as soon as apples and other fruits and vegetables are gathered into boxes, lugs, baskets, or trucks during the harvest, they become contaminated with spoilage organisms from each other and from containers which have not been adequately sanitized.

During the trip to the warehouse or market there will be bumps on the road, railroad tracks to cross . . . perhaps even a wreck on the highway. Mechanical bruising from rough roads or careless handling will encourage the growth of the microorganisms which are ready and waiting to start the decaying process.

As soon as they have arrived at their destination, many vegetables go through a preliminary soaking or agitation in water or a spray treatment. Soaking and washing by agitation tend to distribute spoilage organisms from damaged to whole foods. The washing process can also moisten the surfaces enough to allow organisms to grow and flourish during a later storage period.

The fruits and vegetables which are sold in the retail market without processing, may be contaminated further in the market's bins or other containers, and through contact with decaying products. But the chief offenders in the supermarket are we, the consumers, as we pick up, inspect, squeeze and possibly breathe, cough, or sneeze on the produce.

Microorganisms on the surfaces of freshly harvested fruits and vegetables include not only those of the normal surface variety, but also those from soil and water and maybe even plant pathogens. Any number of molds may also be present, and sometimes a few yeasts.

SPOILAGE

We are only too aware of what vegetable and fruit spoilage is like. If we forget the potatoes for a mite longer than we should, we're confronted one day with a smell that makes it impossible to forget them any longer.

Spinach will turn slimy and dark and foul-smelling in its cellophane sack, and bananas will turn to dark mush before our eyes.

This deterioration of raw vegetables and fruits is caused by physical factors, action of their own enzymes, microbial action, and a combination of all three.

We must realize, too, that the fitness of food for consumption is also judged partly on maturity. If the desired stage of maturity is greatly increased, the food may be considered inedible or even spoiled. One example is our overripe banana, with its black skin and brown, mushy interior.

We need to keep in mind that plant enzymes continue their activity in raw plant foods. If oxygen is available, the plant cells will respire as long as they are alive.

The most common or predominant types of spoilage vary not

KEEPING FOOD SAFE

only with the kind of fruit or vegetable, but also to some extent with the variety. Microbial spoilage may be due to:

(1) Plant pathogens' action on the stems, leaves, flowers, or roots of the plant, on the fruits or other special parts used as foods.
(2) Saprophytic organisms, which may enter a healthy fruit or vegetable as in the case of various rots, or grow on its surface, as when bacteria multiply on moist, piled-up vegetables.

The composition of the fruit or vegetable influences the type of spoilage. So to us, this means bacterial soft rot is likely among vegetables and fruits which are not very acid.

Molds are the most common type of spoilage in fruits and vegetables. Most fresh produce is somewhat acid, fairly dry at the surface and deficient in B vitamins.

Fruits like strawberries, cucumbers, and melons may be in direct contact with the surface of the soil. Spoilage possibilities then rise dramatically. When the fruit is soft and juicy, there is apt to be soft rot and mush. But there are also some kinds of spoilage that have a drying effect so that dry or leathery rots or discolored surfaces may result. In some cases most of the growth of the mold is subsurface, and no rotten spot shows. (Rotting apples are good examples.)

Although molds can and do grow on the surface of fruit and vegetable juices, the high moisture content favors the faster-growing yeasts and bacteria. Which of the two will predominate in juices low in sugar and acid depend more on the temperature than on the composition. The more the temperature drops toward freezing, the more likely is the growth of bacteria and molds rather than yeasts.

In addition to the usual alcoholic fermentation, fruit juices may also undergo other changes caused by microorganisms:

(1) The lactic acid fermentation of sugars.
(2) The fermentation of organic acids.
(3) Slime production.

PROBLEM FOODS AND SAFEGUARDS

Dried fruit (those gems of the orchards which have sustained mankind through centuries of hard, cold winters) pose more problems now than they once did.

For centuries processors were able to dry some fruits in the open air. Farmers themselves used to dry fresh fruits on their roofs in the warm autumn sunlight.

Now, even though we would like to be able to do the same, most of us do not dare, because of the pollution in our urban areas.

Commercial processors and dryers use a sulfur dioxide compound to combat the darkening caused by enzymes.

As a general rule, when we buy dried fruits, we should buy no more than what we think we can reasonably use in a relatively short time. Most will deteriorate in a few months and they are likely to be infested with insects as they begin to age.

SAFEGUARDS

Vegetables

Some of the more dramatic ways to reduce the numbers of microorganisms on foods are by steam peeling, hot water, lye, and blanching.

But most of these safeguards apply to the commercial processor. We should remember that really thorough washing of vegetables (and preferably one at a time) in the home kitchen will remove most of the contaminants.

Unless the wash water is clean it may even add organisms, and then rapid growth of microorganisms can take place on the moist surfaces.

Chlorinated water or borax solutions are sometimes used in commercial establishments for washing, and detergents may even be added to facilitate the removal of dirt and microorganisms. Part of the mold growth on strawberries, for instance, can be removed by washing with a nonionic detergent solution.

We should remember, too, that we must never store bruised vegetables near firm ones.

A few kinds of vegetables that are relatively stable, such as potatoes, cabbage, and celery, can be preserved for a limited time by common cellar storage.

Most vegetables may be kept without special processing but should be cooled promptly and kept chilled. However, each vegetable has its own optimal temperature and relative humidity for chilling storage.

When we deal with vegetables that are to be dried, frozen or canned, we scald them to inactivate their enzymes. At the same time, of course, we appreciably reduce their microorganism count.

Freezing reduces the number of organisms by a percentage that varies with kinds and numbers originally present, but on the average about half of them are killed. During storage in the freezer there is a steady decrease in numbers of organisms. The survivors are ready and waiting to multiply after the storage period.

As the small packages of quick-frozen vegetables usually are placed directly into boiling water and cooked, there is little further opportunity for microbial growth.

When thawed vegetables are held at room temperatures for any considerable period, however, there is a chance that food-poisoning bacteria can grow and produce toxin.

The spores of *Clostridium botulinum,* the dread producer of botulism, have been found in frozen vegetables and can be assumed to be present often. Fortunately, however, the conditions for growth and toxin production would have to be unusual.

Cooking frozen vegetables will not kill all the spores of *Clostridium botulinum,* so foods of this sort should not be allowed to stand at room temperature for extended periods. Boiling thawed frozen vegetables for 15 minutes will destroy any botulism toxin that might have been formed.

Drying by heat destroys yeasts and most bacteria, but spores of bacteria and molds usually survive, as do the more heat-resistant cells. Microbial counts on dried vegetables usually are higher than on dried fruits.

Samples of dried vegetables from retail markets have been found to contain microorganisms in the hundreds of thousands or even millions per gram.

When dried vegetables are sulfured to preserve a light color, their microbial content is reduced. However, if vegetables are dried adequately and stored properly, there will be no growth of microorganisms in them.

The addition of preservatives to fresh vegetables is not common, although the surfaces of some may get special treatment before we see them. Rutabagas and turnips are sometimes paraffined to lengthen their keeping time, and chlorinated water solutions may have been used to wash other kinds.

Sodium chloride is the only added chemical preservative in common use. The amount may vary from 2.25 to 2.5 percent in the making of sauerkraut up to saturation for cauliflower.

Bacteria may cause acid fermentations in foods with lower concentrations of salt. Vegetables that are high in protein and low in carbohydrates, like green peas and lima beans, and some that soften readily, such as onions and cauliflower, are preserved by the addition of enough salt to prevent any fermentation.

Fruits

There are few aromas so appealing as that of apples which are tucked away carefully in a basement or cellar.

But there are very few fruits which can be preserved for any time at all in common cellar storage. Usually, most fruits must be stored in a chilling room where lower temperatures and controlled humidity are used.

A trip to a wholesale produce section in any big city is an eye-opening experience. There are enormous rooms filled with nothing but bananas, oranges, and limes. Each room has its own optimum temperature and its own perfect humidity. The men and women who care for fresh produce are specialists when it comes to determining just how much humidity and warmth are needed in order to ripen fruits.

Frequently, fruit is enclosed in special wrappers and often is treated with chemicals to aid in preservation.

Freezing will reduce the numbers of microorganisms, but it usually causes more damage to fruit tissues than holding in cold storage. Freezing will generally result in flabbiness and a release of juice.

Dried fruits are treated with a sulfur dioxide compound to slow down enzyme action. They should be covered tightly in a cool, dark place, and we must watch them closely for insect infestations as they age.

The safeguards set down in this chapter take the place of the superstitions our ancestors relied upon to avoid contamination and spoilage of foods. If we apply these safeguards and use our common sense these problem foods should no longer trouble us.

TABLE 4

GUIDE FOR STORING
FRUITS AND VEGETABLES (7)

Hold at Room Temperature Until Ripened; Then Refrigerate:
- Apples
- Apricots
- Avocados
- Berries
- Cherries
- Grapes
- Melons (except watermelons)
- Nectarines
- Peaches
- Pears
- Plums
- Tomatoes

Store in Cool Area or Refrigerate, Uncovered:
- Grapefruit
- Lemons
- Limes
- Oranges

Store in Cool Room Away From Bright Light:
- Onions (mature)
- Potatoes
- Rutabagas
- Squash (winter)
- Sweet potatoes

Store in Refrigerator, Covered:
- Asparagus
- Beans (snap or wax)
- Beets
- Broccoli
- Cabbage
- Cucumbers
- Carrots
- Cauliflower
- Celery
- Corn, husked

PROBLEM FOODS AND SAFEGUARDS

    Greens   Peppers (green)
    Onions (green)   Radishes
    Parsnips   Squash (summer)
    Peas, shelled   Turnips

Refrigerate, Uncovered:
    Beans (lima in pods)
    Corn (in husks)
    Peas (in pods)
    Pineapples
    Watermelons

PART 4

Fish and Seafoods

CONTAMINATION

In November 1973, Houston, Texas, city health officials reported at least 187 cases of hepatitis which they later traced to eating contaminated oysters.

Dr. Robert A. MacLean, chief of the city's Communicable Disease Division, told reporters later that the cases probably represented only a fraction of those people actually affected.

Later in the month federal officials confirmed the virus-infected oysters were taken illegally from posted Louisiana waters (2).

This horror story, besides making us sympathetic with the hepatitis victims, also points up a fundamental fact we must keep in mind at all times: the microorganisms in living fish depend upon the microbial content of the waters in which they live.

The numbers of bacteria in slime and on the skin of newly caught ocean fish and sea animals may be as low as 100 and as high as several million per square centimeter.

If we just wash the fish, we greatly reduce the surface count. Bacteria spread through live fish flesh primarily through the gills. Oysters, like those which felled the victims in Houston, pass large amounts of water through their bodies and thus pick

KEEPING FOOD SAFE

up soil and microorganisms. Of course, if pathogens are present in the water, these are picked up too. The same applies to shrimp, crabs, lobsters, and other shellfish.

Of all the flesh foods, fish is the most susceptible to oxidation and hydrolysis of fats and to microbial spoilage. Fish and seafoods must be handled swiftly and with utmost sanitary safeguards.

SPOILAGE

Most of us have never stopped to think about the fact that different types of fish spoil at different rates of speed.

Some flat fish spoil more readily than round fish, because they pass through rigor mortis more rapidly. Rigor mortis is especially important in the preservation of fish, because it retards postmortem autolysis and bacterial decomposition. Any procedure that lengthens rigor mortis lengthens keeping time.

If there isn't a lot of muscular activity before death and if fish are not handled roughly and bruised during catching and processing, spoilage will be delayed. Fish which are exhausted because of struggling and lack of oxygen spoil more rapidly than fish brought in with less ado. Fish which are full of food when caught are also more perishable than those with empty intestinal tracts.

Since the change is gradual from a fresh condition to staleness and then to spoilage, it is difficult to determine the first appearance of spoilage.

But some of the external changes seen as fish spoil are the fading of the bright, characteristic colors and the appearance of dirty, yellow or brown discoloration. The slime on the skin of fish increases, especially at the flaps and gills. The eyes gradually sink and shrink, and the pupil becomes cloudy and the cornea opaque. The gills turn a light pink and finally grayish yellow in color.

The flesh softens and exudes juice when it is squeezed, and we can easily indent it with our fingers. The normal fresh, seaweedy odor changes first to a sickly sweet one, then a stale-fishy odor, and finally a putrid odor. Cooking will bring out the odors more strongly.

Some fish have a musty odor and taste which is caused by

PROBLEM FOODS AND SAFEGUARDS

Streptomyces growing in the mud at the bottom of the lake or river. The putrid yellow and yellowish green colors that are so repulsive-looking in spoiled fish are caused by *Pseudomonas fluorescens; Sarcina* and *Bacillus* species cause the red or pink colors of spoilage.

When we smoke fish, the chief spoilage organisms we have to worry about are molds. Marinated fish should pose no real problem unless the acid content of the marinade is low enough to let bacterial growth flourish or unless there is enough air in the container to permit mold to grow.

Oysters, particularly when they are shucked, deteriorate and spoil very rapidly. That is why oysters are still in their shells on ice in fish markets.

SAFEGUARDS

Since fish are so perishable, they must be refrigerated. But man has not always been able to buy a red snapper at the corner fish market and go right home and put it in the refrigerator for tonight's dinner.

The first crude attempts at refrigeration utilized snow banks in the far north. The people who lived in the near-arctic areas were lucky as far as food preservation went.

Back in 1838, a burly captain of a Gloucester, Massachusetts, smack had the idea of putting ice on board to preserve a catch of halibut. By 1861, fish were being frozen with ice and salt almost routinely in some areas.

But the ice used to preserve fish in 1861 and the ice used today by a modern fish processing ship differ greatly. We know today, for instance, that ice made with contaminated water is little better than no ice at all.

Canning is another successful (perhaps the most successful) method of preserving fish. However, fish, like meat, is low in acid and has a slow penetration rate for heat, so great care must be taken in canning procedures.

Oysters are often packed into unsealed cans and not heat processed, but they must be refrigerated, and they must be used very quickly.

KEEPING FOOD SAFE

Fish have been packed in brine since the early days of Rome. (Salt controls bacterial growth very efficiently.) Dried fish were a staple in early Egypt. But since our modern palates seem to prefer the fresh or frozen forms of fish, we need to direct our attention to controlling the growth of microorganisms before processing.

Safe keeping of fish and seafood varies with the variety but, in general, it should not be kept more than one or two days.

Right after catching, fresh water fish should be stored in clean containers and kept well chilled. As soon as possible, they should be washed thoroughly, gutted, and cleaned. Fish should be frozen immediately if it is not to be eaten within a day or two.

Leftover cooked fish should not be kept more than one or two days.

PART 5

Eggs

CONTAMINATION

Modern man has so lost touch with the farm and farming that he is blissfully ignorant of the common, everyday sources of contamination for eggs.

Eggs may have been contaminated by fecal matter from the hen or the lining of the nest. We should clean the egg storage area in our refrigerators more often.

Eggs can also be contaminated from the wash water or by the boxes they are packed in.

Salmonella species may be on the shell or in the egg itself when it is laid. The organisms can build up during processing and they can be present in very significant numbers in frozen or dried eggs.

SPOILAGE

When it comes to washing eggs, egg processors are damned if they do and damned if they don't. Hand-washed eggs are more likely to rot than unwashed ones, and machine-washed eggs provide more types of rots than hand-washed ones.

PROBLEM FOODS AND SAFEGUARDS

Washing with warm, plain water removes the bloom (the thin surface layer of proteinaceous material on the shell) as well as some of the microorganisms, but it also encourages the penetration of bacteria right through the pores in the shell.

Unless precautions are taken, the wash water will build up great numbers of microorganisms.

Untreated eggs lose moisture during storage and thus lose weight. The amount of shrinkage is shown to the candler by the size of the air space at the blunt end of the egg. A large area indicates a lot of shrinkage.

The egg white becomes thinner and more watery as the egg ages. The yolk becomes weaker and may even break when the shell is opened. We have all noticed that some eggs will flatten out in a frypan. This indicates an old egg; a fresh egg has a thick white and a yolk which stands up in the form of a hemisphere.

But as we know, age is not the only spoiler we have to worry about.

Microorganisms can contaminate the shell of an egg and penetrate the pores to the shell membrane. (Usually the shell must be moist for this to happen.) The organisms grow through the shell membranes to reach the white. After they grow in the white they reach the yolk where they can grow very rapidly and complete the spoilage of the egg.

The time required for bacteria to penetrate the shell membranes varies with the organism and the temperature, but they may take as long as several weeks at common refrigerator temperatures.

(Those of us unfortunate enough to have smelled a rotten egg may have to force ourselves to go on at this point.) Nevertheless, black rot or—even worse to contemplate—red rot means the egg at some point has been held at relatively high temperatures, higher than those usually used for storage.

Eggs can also be spoiled by mold fungi. Pin-spot molding of eggs is actually the first stage of fungal growth. The colonies of molds are small and compact and occur both on the shell and just inside it. The colors can vary from yellow to blue or green. *Penicillium* causes yellow and green or blue spots. A variety of *Sporotrichum* causes pink spots.

If eggs are stored for long periods of time under very humid conditions, a fuzz will cover the shell and eventually result in a quite luxuriant fungal growth. If the fungus is left to grow, the egg will reach the final spoilage stage after the mold has grown completely through the pores or cracks in the egg. The egg white may jell and produce colored rots. Even the yolk membrane may rupture.

Sometimes spoilage causes little more than an off flavor with very little other evidence of damage. Bacteria may cause a sort of musty aroma, and eggs which have been allowed to stay on straw for longer than usual may develop an earthy or strawy flavor which is caused by *Streptomyces* on the straw and is later absorbed into the egg itself. Once in a while an egg will have a fragrant aroma of hay; this is caused by *Citnobacter cloacae*. Some strains of *Escherichia coli* will produce fishy odors and flavors.

The well-known "cold-storage" taste is often absorbed from packing materials.

SAFEGUARDS

Three basic rules govern how we take care of eggs. They must be kept dry, cool, and covered.

Eggs have a natural protection (the outer and inner membranes inside the shell) which holds back bacteria.

Eggs should not be held for lengthy periods, because aging allows changes to occur in their membranes which favor rapid growth of bacteria.

They should never be washed before storage unless they are washed in a warm detergent sanitizer solution. (We should take care not to buy extremely dirty eggs in the first place.)

An excellent treatment for long-term storage eggs is immersion in a hot detergent-sanitizer solution (at 110° to 130° F.). The sanitizer should be a quaternary ammonium compound.

A new thermostabilization method of dipping eggs into hot water reduces the evaporation of moisture by an extremely slight coagulation of the outer layer of albumin.

The Code of Federal Regulations recommends that all liquid eggs, except whites, be pasteurized at 140° F. for not less than

PROBLEM FOODS AND SAFEGUARDS

3.5 minutes. The code also specifies that eggs must be free from salmonellae under tests conducted in the laboratory. Frozen whites, according to the code, should be dehydrated in such a manner that they too will be free of salmonellae.

On January 1, 1966, it became compulsory to pasteurize liquid whole eggs. Liquid whites of eggs fell under the same regulations on June 1, 1966.

Sometimes the shells are treated with a colorless and odorless mineral oil.

The measures used to improve keeping qualities during cold storage are:

(1) Keeping the shell surface dry.
(2) Preventing loss of moisture.
(3) Cutting down on the circulation of air.
(4) A general slowing down or stopping of any changes due to microorganisms or enzymes.

Eggs are selected for cold storage on the basis of their appearance and the results of candling. When an egg is candled, it is held and turned in front of a light so it can be examined for cracks, rots, molds, blood, large air cells, or developing embryos. (That's why we seldom find a half-developed chicken in a commercial egg.)

After candling, eggs are cooled quickly and held under a humidity condition that depends upon the length of storage. If the humidity is too low, the eggs will lose moisture and weight. If the humidity is too high, microbial spoilage is likely to occur.

The higher the temperature is above 29° F., the more readily microorganisms will penetrate the shells.

Eggs have always been tempting to mankind. They are a ready source of nourishment, but that old spoilage problem has always been a primary concern.

The ancient Chinese got around the problem of preserving eggs by storing them in large casks for 35 to 60 hours. The method utilized a natural fermentation of the egg white by coliform bacteria. There were large numbers of bacteria in the egg white both before and after drying.

Now we know that dried eggs may contain from a few hundred microorganisms per gram up to over a hundred million, depending upon the methods of handling. But the drying process can reduce the microbial content a hundredfold from the numbers in the liquid egg. Adequately dried egg products contain too little moisture for growth of microorganisms.

For many, many years preservatives of all kinds have been used on the shells to ward off spoilage. Eggs have been stored fairly successfully in salt, lime, sand, sawdust, and ashes. Immersion in a solution of sodium silicate has been a time-honored method of preservation.

Because eggs are porous, they will lose flavor and moisture if stored uncovered. Egg cartons provide excellent storage conditions.

We should not use cracked eggs in foods which are not thoroughly cooked. (This applies particularly to eggnogs, milk shakes, soft-boiled or scrambled eggs, ice creams, and meringues.) But cracked eggs shouldn't be thrown out. They can be used safely in dishes which are well cooked in the oven or on the surface of the range. Dried or frozen eggs should be used only in foods which will be thoroughly cooked unless the label has said the eggs are safe for partially cooked or uncooked dishes.

Leftover yolks should be covered with cold water and stored in the refrigerator in a covered container. Extra egg whites should also be stored covered in the refrigerator. Both should be used within 2 to 4 days.

Dried eggs should also be kept in the refrigerator. After a package has been opened, the unused portion should be stored in an airtight container.

Dried egg will keep its good flavor for about a year if it is stored properly.

Because eggs may be infected with *Salmonella* organisms, we should wash our hands with soap and water after handling them (22).

And whether we like it or not, we must realize that when we eat raw eggs or products containing raw eggs, we are running the risk of taking *Salmonella* organisms into our own bodies.

We must be reasonable, however, and realize we may be exposed to *Salmonella* day in and day out. It is only when we allow conditions to exist which favor rapid growth and development of the organisms that we are likely to become ill.

PART 6

Poultry

CONTAMINATION

Poultry which has not been properly cleaned will develop off flavors and a high microbial count. The bacteria and molds which occur naturally on the skin of fowl contribute their share of contamination. Add to that the even heavier contamination from feathers and feet during plucking and washing and we have a potential contamination problem of immense proportions.

The type of feed the fowl has consumed just before slaughter can have an influence on any visceral taints.

Contamination can occur during production, handling, and storage in manufacturing plants. But it can also occur right in our own kitchens. Many of us are unknowing carriers of organisms like *Salmonella*.

Figures compiled by the Center for Communicable Diseases indicate that at least 25,000 people annually may shed salmonellae into the food chain. (Salmonellae are a major problem in poultry.)

SPOILAGE

Enzymes in fowl lead to the deterioration of dressed birds, but the chief and, by far, most important spoilers are bacteria. The bird's intestines are the primary source. Most bacterial growth takes place on the skin, the lining of the body cavity, and any surfaces which have been cut. Decomposition spreads slowly from these surfaces right on into the meat.

Salmonella, claims the USDA, is the organism which is most likely to be found in fowl. In fact, we should assume that

KEEPING FOOD SAFE

Salmonella organisms are present in all uncooked poultry (17).

Chilling, freezing, and thorough cooking stop the growth of *Salmonella* organisms temporarily, but they will start growing and reproducing again if the fowl is allowed to remain at dangerous temperatures (40 to 140° F.).

Unfortunately, salmonellae usually do not affect the appearance, smell, or taste of food.

SAFEGUARDS

The Federal Government is well aware of the dangers of salmonellae and other organisms.

In July 1973 the United States Department of Agriculture announced a major campaign to control and eliminate *Salmonella* throughout the food chain.

The campaign involves the expansion and coordination of intensive consumer education aimed at eliminating careless food handling practices in the home and food service establishment. It emphasizes:

(1) Continuation of a campaign to eliminate salmonellae from rendered animal by-products used in animal foods.
(2) Modification of processing procedures and facilities in meat and poultry plants.
(3) Intensified support of research aimed at eliminating *Salmonella*.
(4) The development of model ordinances governing sanitation and food handling in retail stores and food service institutions.
(5) Standards for the food transportation industry.

Clearly, the problem of safeguarding our food from salmonellae is not a simple one.

Until about twenty years ago, meat and poultry inspectors relied almost entirely upon their senses of sight, touch, and smell to detect abnormalities. Now, however, federal detection methods can pinpoint microorganisms in animal tissue in terms of parts per billion or even parts per trillion.

PROBLEM FOODS AND SAFEGUARDS

Mistakes we make in our kitchens, restaurants, food retail establishments, and institutions account for about 80 percent of all foodborne diseases and about 53 percent of salmonellosis outbreaks (22).

However, the *Salmonella* chain can be broken before the organisms have a chance to grow in our kitchens, our foods, and our bodies by following these safeguards:

(1) Wash poultry carefully and thoroughly.
(2) Wash hands thoroughly with hot water and soap after handling raw poultry to lessen the possibility of contaminating cooked, ready-to-eat foods or serving utensils.
(3) Store poultry for no more than two days in refrigerator at or below 40° F.
(4) Thoroughly cook poultry at low to medium heat (300 to 350° F.) to an internal temperature of 170° F. for chicken, 185° F. for turkey, 175° F. for boneless turkey roasts, and 165° F. for poultry stuffing.
(5) Do not thaw frozen raw poultry at room temperature. It should be thawed in the refrigerator or under cool, running water in a waterproof container.
(6) Do not stuff chickens or turkeys the night before cooking.
(7) Leftover poultry gravies should be reheated to a rolling boil before serving.
(8) All utensils and surfaces used in preparing raw poultry should be washed with hot water and detergent before using again.
(9) Never eat raw or partly cooked poultry.
(10) Cook poultry immediately after thawing.
(11) Never partially cook poultry and then finish the job later on.

Poultry should be frozen if we do not intend to cook and serve it within one to two days.

Refrigerator temperatures between 35° and 40° F. should be maintained so the *Salmonella* organisms cannot grow.

We should keep poultry in the coldest part of the refrigerator, which is generally near the ice cube compartment or in a special meatkeeper.

The special wrap on prepackaged products can be left on if we use the poultry within the safe period (one to two days). Raw poultry purchased in a meat market should be unwrapped and then rewrapped in wax or plastic paper and placed on a plate or tray so it cannot touch any other food in the refrigerator. Giblets should be wrapped and stored separately.

We should also make sure that thawing poultry cannot touch other foods. It should thaw in a container which will not allow blood or moisture from the poultry to drip on or touch anything else.

PART 7

Milk and Milk Products

CONTAMINATION

When milk comes from the udder of a healthy cow, it contains relatively few bacteria. Once milk leaves the udder it is subject to contamination from many sources. Bacteria can come from the air in the barn, milk cans, flies, pipelines, milk coolers, etc.

And any of the dairy products made from milk may be subject to even further contamination.

SPOILAGE

Because milk is high in moisture, nearly neutral in pH, and rich in microbial nourishment, many kinds of microorganisms flourish.

The chief types of spoilage are:

(1) Souring or acid formation.
(2) Gas production.
(3) Proteolysis (an enzyme transformation).
(4) Ropiness.

PROBLEM FOODS AND SAFEGUARDS

(5) Changes in butterfat.
(6) Alkali production.
(7) Flavor changes.
(8) Color changes.

When milk sours, we tend to think of it as being spoiled (especially if it curdles). However, the lactic acid fermentation of milk is used to make cheese, yogurt, and buttermilk. (Curds and whey have a respected place in history, to say nothing of nursery rhymes.) This lactic acid fermentation is most likely to happen at room temperature.

When acid fermentation occurs we first notice a sour taste, then coagulation. (Pasteurization kills the more active acid-forming bacteria, but will permit heat-resistant lactics to survive.)

The chief gas-formers are coliform bacteria, *Clostridium* species, the aerobacilli (which produce both hydrogen and carbon dioxide), yeasts, propionics and lactics.

We know milk has gas in it if:

(1) there is foam at the top of the milk,
(2) gas bubbles are caught in the curd or furrows, and
(3) there are floating, curd-containing gas bubbles.

Yeasts are usually absent or in very low numbers in milk and generally cannot even compete with the bacteria.

In proteolysis a bitter flavor is usually caused by some of the peptides. This type of spoilage is favored by low temperature storage and molds and film yeasts which destroy acids. Acid proteolysis (perhaps like Miss Muffet's) causes a shrunken curd and a lot of whey.

Ropiness or sliminess can be found in milk, cream, or whey. Non-bacterial scum or skin can be caused by:

(1) Mastitis from the cow.
(2) Thickness of cream.
(3) Thin film of casein or lactalbumin during cooling.

Bacterial ropiness is caused by slimy capsular material and generally is favored by cool temperatures.

Butterfat can be spoiled by many bacteria, yeasts, and molds which cause off odors and tastes and rancidity.

Milk drawn straight from the cow and tasted immediately may have an off odor and off taste because of mastitis, stage of lactation of the cow, or the type of feed consumed.

No one has to tell us when milk is spoiling, because we can taste the acid or bitter, sometimes burnt, musty or soapy flavors. Sometimes, spoiled milk even has a fishy taste.

Even though pasteurization kills most of the microorganisms, enough re-enter after heating to cause a slow formation of bitterness and other off flavors. Many types of bacteria grow at room temperature.

Buttermilk may spoil from mold growth when the surface is exposed to the air. Otherwise, buttermilk keeps for lengthy periods of time, because the acid content is too great to allow growth of most microorganisms.

Dry milk is so low in moisture that little or no microbial spoilage occurs when it is handled properly.

Unsweetened evaporated milk is canned and heat-processed under steam pressure in order to destroy all microorganisms. Spoilage can occur, however, if the heat process is not adequate or if the can is defective.

Even in this highly processed form of milk, the spores are waiting for the right time and place. Under certain conditions, they may cause cans to swell or the milk to coagulate and take on a bitter flavor.

Spoilage of sweetened condensed milk is due primarily to organisms which have entered after the heat treatments, especially through the air.

Ice cream, ice milk, frozen custards, sherbets, and ices do not spoil easily, because the mix is pasteurized before it is frozen. However, if they are held at temperatures above freezing, souring from acid-forming bacteria may occur.

In this country, most butter is made from pasteurized cream, and butter is generally kept in the refrigerator, so it is unusual for it to spoil.

Soft cheeses like Limburger and Brie are quite perishable, because of their high moisture content. Hard cheeses like Cheddar and Swiss are the most stable.

We are all familiar with the molds that grow on the surfaces of cheese. Mankind has been scraping mold off cheese for centuries.

Cheese can be traced back through history to the times when the Arabs carried milk in an animal pouch. (These pouches were generally goat stomachs, which had rennet occurring naturally in the lining.)

After miles of jiggling on a camel back in the heat of the desert, our Arabs soon had curds and whey. Cheese was the next logical step.

And think of the mold problem Queen Victoria must have had when she received a 1,100-pound package of cheese for a wedding gift when she married Prince Albert!

SAFEGUARDS

The problem with guarding milk against spoilage has always been one of taste. No rigorous preservative methods could be used, because the delicate taste would be affected. That is why pasteurization was such a historic innovation.

Pasteurization involves one of several legalized heat treatments. They are: 145° F. for thirty minutes; 161° F. for fifteen seconds, or any other procedure shown to be as effective as these treatments. The milk is then cooled immediately to 45° F. or lower.

Pasteurization helps prevent diseases which can be spread from cattle to humans.

The purposes of pasteurization are:

(1) To kill all pathogens that might be in the milk.
(2) To improve the keeping qualities of the milk.
(3) To destroy microorganisms which would interfere with the activities of desirable organisms (such as starter bacteria) or that would cause spoilage of butter or cheese. Pasteurization should kill most of the microorganisms in milk.

But pasteurization cannot make unclean milk clean, only safer. Dairy utensils must be adequately cleansed and sanitized. Cows must be kept well groomed, and the udders should be cleansed with a germicide before milking. Milking areas should be paved and washed frequently with germicidal solutions.

Today, there is an equally significant development in processing milk—sterilization. After sterilization, no refrigeration is needed during distribution or storage.

Technology has at last made it possible to sterilize milk by heating it to about 280° to 300° F. for about three to four seconds without any rather objectionable results.

But until economic factors make it feasible to sterilize milk on a mass-production basis, we must still be very careful to keep our milk stored at about 40° F.

Recent research has shown that lowering storage temperatures to just above freezing can extend the life of milk up to seven weeks (20).

Besides storing milk at about 40° F. in our home refrigerators, we should use it within ten days. Cream, butter, and cheeses should also be kept in the refrigerator except when we are using them.

Milk and cream should be kept covered so they will not take up odors and flavors from other foods.

There was a time when very few people used dry milk, but more and more of us are finding it a practical way of keeping a steady supply of milk on hand at all times.

We should keep our dry milk (either nonfat or whole) in tightly closed containers away from heat. Nonfat dry milk will keep under good conditions for several months in our cupboards. But we should be careful to always close the container tightly after we have used it, because lumps and a stale taste become problems if dry milk is exposed to the air for any length of time.

Dry whole milk is generally sold only for infant feeding; it doesn't keep as well or as long as nonfat milk because of its high-fat content. Once a container of dried whole milk has been opened, it should be stored, tightly covered, in the refrigerator.

When we reconstitute dry whole milk or nonfat dry milk, we should store them in the refrigerator and treat both as though they were fresh, fluid milks. Chilling makes the flavor of both distinctly better.

PROBLEM FOODS AND SAFEGUARDS

Those handy little cans of evaporated and condensed milk are some of the soundest buys we can make if we want to keep milk on hand for instant use at any time. They should be stored at room temperatures until we open them; then we should cover tightly and refrigerate the milks as though they were fresh, fluid milk.

Cheese foods and cheese spreads have to be wrapped tightly or covered after opening, and must be stored in the refrigerator.

If cheese is brushed lightly with vinegar before wrapping or is wrapped in a vinegar-soaked cloth, the chance of mold growing on the cheese is lowered.

Both natural and hard cheeses should be kept tightly wrapped and refrigerated. The cut surfaces of large pieces of hard cheese can be dipped in hot paraffin to help keep them from drying out. (They can also be frozen, but the freezing can cause crumbling.)

Cottage cheese must be used within three to five days, but other soft cheeses like cream or Camembert can be kept up to two weeks. In general, a good rule to follow in handling dairy products in our kitchens is to keep them cool, clean, and covered.

PART 8

Meat and Meat Products

CONTAMINATION

On November 1, 1973, a check of ground beef bought in supermarkets in several major U.S. cities determined that the meat was frequently contaminated with fecal bacteria (3).

Because of the many sources of contamination, the types of bacteria, molds, and insects which can contaminate meat are varied. There is also the possibility of meat and meat products being contaminated with human pathogens, especially those of the intestinal type.

The healthy inner flesh of meats generally does not harbor microorganisms, but they have been found in lymph nodes and bone marrow. The significant contamination comes from external sources during bleeding, handling, and processing.

SPOILAGE

That piece of raw animal flesh our cave man hid under the rock when the rest of the tribe wasn't looking was just the beginning of our spoilage problems with meat.

Raw, fresh meat is changed drastically by its own enzymes as well as by the chemical oxidation of its fats. A very moderate amount of autolysis is desirable when we tenderize beef or game by hanging or aging, but it's the last thing we want in most other raw meats. (Autolysis is a type of self-digestion of meat after it has been butchered.)

But we're asking for real trouble if we try to age meat at home, because we cannot maintain constant temperature and humidity, so the numbers and kinds of bacteria are almost impossible to control.

There are four basic rules which should be kept in mind whenever animals are slaughtered:

(1) Animals should be starved for 24 hours just prior to slaughter to reduce the bacterial load in the intestinal tract.
(2) The more rapid and sanitary the slaughter conditions have been, the fewer organisms will be invading the meat.
(3) Rapid cooling after slaughter will cut down considerably on the growth of bacteria.
(4) Bacteria are more likely to become major problems if the animal is fatigued or excited just before slaughter. (For example, these conditions may prevent complete bleeding, which in turn may affect bacterial growth and chemical changes.)

These rules also apply to wild game. When we are dealing with wild game, food safety must be utmost in our minds. It is essential that an animal be field dressed, cooled, and skinned as soon as possible, and that cool temperatures must be maintained throughout the entire aging period. (It is best to hang an animal off the ground to insure adequate air circulation.)

PROBLEM FOODS AND SAFEGUARDS

Here again, the danger zone between 40° and 120° F. must be kept in mind. We should drive home from the hunt during the cool, nighttime hours; however, if we must drive during the day, when it is warm and sunny, the animal should be insulated.

The common types of spoilage we are likely to encounter in meat are surface slimes, changes in color of the meat, phosphorescence, changes in fat, odors and tastes, souring, putrefaction, and the growth of mold. Sometimes meat will spoil only in spots, particularly in the beginning stages of spoilage.

Most meats are perfect hosts for the growth of microorganisms. They are high in moisture, fairly low in acid, and high in microbial food. Add to that the fact that some organisms may already be in the lymph nodes, bones, and muscle. The result: contamination with one or more of the spoilers is almost impossible to avoid. These factors make meat one of the most difficult kinds of food to preserve.

SAFEGUARDS

If it weren't for federal inspection regulations, we would have even more reason to be concerned about meat than we do. But fortunately, federal inspection standards are rigid. In addition, many states and cities have their own inspection laws as well.

All meat and meat products prepared in plants which ship meat into other states must be federally inspected. (In this country, "meat" means the "red meats" from cattle, sheep, and hogs.) This should assure us that the meat comes from healthy animals, is processed under sanitary conditions, is properly labeled and packaged, and that it is not adulterated or contaminated. A round stamp identifies meat which has passed this inspection:

(This stamp should not be confused with the shield-shaped stamp which specifies grades.) States also use their own stamps to indicate inspection.

Pork occasionally contains trichinae (tiny parasites) which cause a disease in humans called trichinosis. (See Chapter 1.)

Commercially, pork can be frozen at temperatures not possible in the home kitchen. This also greatly reduces the possibility of trichinosis. But when it comes to cooking pork, the only practical real protection we have is what we provide ourselves . . . by thorough cooking of all fresh pork.

All fresh pork should be cooked until the meat is no longer pink. This means an internal temperature of 170° F. for loin roasts and 180° F. for all other roasts, such as fresh hams.

There is another method of killing the organisms. This is by freezing, but temperatures should be checked with an accurate freezer thermometer.

The pork should be cut up into separate pieces no larger than 6 inches in depth, thickness, or width.

There are three rules which should be observed to the letter if this method is used:

(1) At temperatures of 5° F. in a home freezer, the pork should be kept frozen 20 days.
(2) At temperatures of −10° F. the pork should be kept frozen 10 days.
(3) At temperatures of −20° F. the pork should be kept frozen 6 days.

Inspection of meat actually begins in the holding pen before slaughter in order to identify and eliminate any diseased animals from the food channel.

Inspection of carcasses and internal organs is made by veterinarians and their assistants who have had special training.

In addition, a U. S. Department of Agriculture inspector examines the equipment in a meat plant to make sure it is clean before work begins each day. Then he checks all the operations in the slaughtering or processing plant to make sure products are handled properly and equipment is cleaned as often as necessary.

In 1972 alone, 871 tons of meat foods were detained in trade channels in seven Southwest states by compliance officers of the U. S. Department of Agriculture's Animal and Plant Health Inspection service. (Most of this was detained for suspected adulteration, mislabeling, or for a violation of meat inspection regulations.) (16)

Unless cooling is prompt and rapid after slaughter, meat may change drastically in appearance and flavor, to say nothing of supporting and favoring the growth of microorganisms before it reaches the processing plant. In addition, under certain circumstances, lengthy storing at even chilling temperatures may result in higher microbial counts.

The most significant safeguard taken in the entire processing and slaughtering chain is that of keeping microorganisms away from meat as much as possible. Once meat is contaminated, it is very difficult to remove microorganisms.

Use of Heat

Canning procedures vary considerably with the type of meat to be preserved. Since most meat and meat products are very low in acid, bacteria which might survive the canning process are quite likely to grow rapidly and spoil the meat.

Chemicals which are added in the curing process usually make the canning procedures more effective.

Bulletins on the home canning of meats are available from federal and state agencies, giving safe processing times and temperatures.

The use of a pressure cooker is mandatory, and meats are generally precooked to make the packing job easier.

Precooking or tenderizing of hams also helps reduce microbial numbers to some extent.

We also need to keep cooked sausages, like frankfurters, and uncooked sausage refrigerated.

Use of Low Temperatures

Chilling temperatures from 29.5° to 36° F. are best. (The lower temperatures are preferable, however.) The time limit for storing beef at these cold temperatures is about thirty days, for pork, lamb, and mutton only one to two weeks, and for veal an even shorter period.

Most meats sold in retail stores have not been frozen, but freezing often is used to preserve meats during shipment over long distances or for holding until times of shortage.

The preservation of frozen meats is increasingly effective as the temperatures drop from 10° toward −20° F. The freezing kills about half the bacteria, and the numbers decrease slowly during storage.

Salmonellae can survive freezing for months at storage temperatures and start growing again at thawing temperatures as low as 38° F.

Drying

In pioneer America, jerky (sun-dried strips of beef) was accepted as good food on a regular basis, particularly on the trail.

Freeze-drying, a new version of the ancient method, is on the increase. And though proper drying tends to reduce the number of microorganisms, no meat should ever be dried which is not low in microbial count to begin with.

Use of Preservatives

Meat has long been one of the most difficult of all foods to keep. For centuries, salt or brine was the only way most people could keep fresh meat over the winter. The meat either had to be rinsed repeatedly before it was used or combined with other foods which would readily take up the excess salt.

Today, the storage life of meat has been extended greatly by the use of more sophisticated preservatives.

The use of controlled atmospheres with added carbon dioxide is a major help. Today, ordinary salting is combined with curing and smoking in order to produce an acceptable product.

Safeguards in the Home

After all of this care and concern on the federal, state, and local level, meat can still spoil right in our own kitchens and become completely unfit for human or animal consumption.

PROBLEM FOODS AND SAFEGUARDS

When we bring perishable meat home from the supermarket, we should store it at once in the refrigerator or freezer. We should always be certain our refrigerators are chilling below 40° F., and our freezers must maintain temperatures at 0° F. or lower.

Raw meat should be loosely wrapped, or if it is completely encased in fat, it should be left uncovered. (Some air drying during storage is good, because a dry surface inhibits the growth of bacteria.) This is why meat should never be washed before it is stored.

If we purchase meat in a small, independent meat market, or if we have the meat specially cut in a supermarket and we bring it home wrapped in paper, all we need do is loosen the ends of the paper. This will give plenty of ventilation. (As a general rule, the larger the piece of meat, the longer we can keep it safely in the refrigerator before cooking it.)

Ground meat, fresh sausage, and the organ meats are the most likely targets for spoilage. All of these should be used within 24 hours after we buy them.

Just a little tender loving care will pay off when we handle ground beef. We should take off the plastic film wrapping and wrap it loosely with wax paper.

Uncooked diced and cubed pieces of meat should never be kept longer than 2 to 3 days.

The reasoning behind these safety rules is that both ground and diced or cubed meat have more surface area exposed where bacteria can grow.

Roasts, however, will keep safely from 3 to 5 days in the refrigerator, and steaks and chops can be kept up to 4 days.

The other meats (lamb, veal, and pork) should not be kept quite as long as beef.

A small book could be written about the care and cooking of hams. Yet how many of us have ever actually read much about either fresh or cured ham storage and keeping qualities?

Hams which have been tenderized by precooking can be safely kept in the refrigerator for two weeks, if the seal of the wrapper has not been broken.

Whole hams should be used within a week, half hams and slices within 3 to 5 days. Mild-cured hams are similar to fresh meats as far as storage is concerned.

With the exception of small (1½- to 3-pound) canned hams, which are not labeled "Perishable, Keep Under Refrigeration," and dry-cured hams such as country-style and Smithfield, ALL ham must be kept refrigerated before cooking.

Unopened canned hams labeled "Perishable, Keep Under Refrigeration" can be kept in the refrigerator up to 6 months without loss of quality.

Sometimes when we see a real bargain in hams, we are tempted to buy several and freeze them. But this is not a good idea, because there are likely to be changes in flavor and texture in hams during freezing. When it is absolutely necessary, cured hams can be frozen for 1 or 2 months if they are wrapped tightly, sealed well in moisture-resistant wrap, and stored at 0° F. or lower.

When we cook and prepare hams, proper sanitation is absolutely essential. The spoiler which is lurking at our elbows in the kitchen when we prepare hams is *Staphylococcus*.

According to 1971 statistics on foodborne illnesses, ham was the meat most often involved in cases of "Staph" food poisoning (18).

When it comes to cooking hams, there are three temperatures to keep in mind, depending upon the kind of ham we select:

(1) If the ham is a canned or fully cooked type, it may be eaten cold without any kind of advance preparation. If we prefer it warm, it can be heated to an internal temperature of 140° F.

(2) If the ham is the cook-before-eating kind, it should be baked or simmered to an internal temperature of 160° F.

(3) If we are cooking shoulder cuts of pork like cured picnics or shoulder butts, they should be heated to an internal temperature of 170° F. Fresh ham, an uncured product, should be treated like fresh pork and heated to 170° F. internal temperature.

PROBLEM FOODS AND SAFEGUARDS

In summing up, there are a number of safeguards we should remember when we handle meat in our kitchens:

(1) Cook meats promptly after thawing.
(2) Cook meat and meat products at low-to-moderate temperatures to assure complete cooking throughout.
(3) Always cook all pork until there is no pink color in either the meat or the juices.
(4) Use internal temperatures as the best guide to the doneness of meats. (See Table 2.)
(5) Serve meats and gravies immediately after cooking; never let them set out for lengthy periods.
(6) Follow instructions for cooking unprocessed meats.
(7) Reheat leftovers thoroughly.
(8) Select meat and meat products just before leaving the store, get them home quickly and refrigerate them.
(9) Keep meat and meat products in the coldest part of the refrigerator.
(10) Keep frozen meat at 0° F. or lower.
(11) Defrost meats in the refrigerator whenever possible.
(12) When meat must be thawed at room temperature, it should be put in a closed, double-thickness paper bag. An even better method is to put the meat in a waterproof container and immerse in cold water.
(13) Do not buy torn or damaged packages of meat when shopping.
(14) Wrap meats correctly for storage . . . loosely for the refrigerator and tightly for freezing.
(15) Thoroughly sanitize all equipment that has been used in the handling of raw meat.
(16) Wash hands with hot water and soap after handling raw meat.

If we remember and apply these safeguards, we can be assured of more wholesome food.

5

Clean and Healthy Kitchens

It is strange indeed that, in the twentieth century, many Americans are more concerned about getting rid of a bad odor than they are about killing potentially dangerous microorganisms on the cutting board.

Television, radio, and the printed media have done much to give us false standards of cleanliness in our homes. Modern technology coupled with a concern about status rather than genuine cleanliness have contributed to an incredible amount of ignorance and complacency on our part.

To further complicate our dilemma, technology has created a myriad of different types of cleansers, plus complicated appliances and dozens of types of surfaces.

We are bombarded by commercials drilling into us the importance of shine, or pine aroma, or labor-saving devices. How many of us realize we must clean our kitchens for one main reason: we must kill or inactivate the dangerous microorganisms in our kitchens.

The Common Sense of Kitchen Cleanliness

When a family member comes home with an infectious disease and remains sick in the house for a day or two, almost any imaginable spot in the kitchen can become contaminated. Furthermore, any one of us can be an innocent and unknowing carrier of disease-causing microorganisms.

But we are not helpless. Far from it. Even though technology has made our lives more complicated, it has also provided us with effective ways to clean and sanitize our kitchens IF we understand and know how to use them.

Water and a readily available compound called elbow grease are still the most important cleansers of all. Water will dissolve more substances than any other liquid. (Of course, soft water is best, since neither soap nor soil will dissolve very easily in hard water.)

Technology has pretty well taken care of the hard water problem, too. Even if we live in a hard-water area, and even if we do not own a water-softening appliance, we still can fall back on the time-honored softeners: borax, ammonia, potash, or lye.

As we have learned, the conditions necessary for the growth of most molds, yeasts, and bacteria are dark, damp, and unclean places plus food and warmth. The most efficient way to clean, therefore, is to remove microorganisms and food with soap and/or detergent, water, and a disinfectant. (A disinfectant is an agent which kills all bacteria.)

There are two general types of disinfectants:

(1) Natural (sunlight and heat).
(2) Chemical (acids and alkalies, phenols, etc.).

The ordinary airing out and sunning which was so popular a generation ago can go a long way toward disrupting some of the conditions which favor the growth of microorganisms. We know that sunlight does kill some microorganisms, and that drying removes the moisture necessary for continued microbial existence.

Alkalies are extremely effective in the removal of grease. Soap, for instance, is a product made by the chemical reaction between an alkali and animal fat, fatty acids or plant oils.

And although soap is not in itself a disinfectant, it frequently contains one. Soap is an efficient cleanser because when used with

CLEAN AND HEALTHY KITCHENS

water it dissolves and unites with any grease it touches. It loosens and washes away the dust and dirt. With the dust and dirt also go the microorganisms.

A detergent is one or more chemicals which remove soils by wetting and penetrating surfaces and by holding the dirt in suspension.

When someone in the family comes down with a contagious illness, we should disinfect the kitchen often. Disinfecting lessens the chances of secondary infections from *Staphylococcus* and *Streptococcus,* to say nothing of *Salmonella* and other disease-causing organisms.

When using a disinfectant we must read and follow the directions and precautions carefully.

As a rule, the protection of disinfectants will last from 48 hours to 7 days, depending upon the type used and how accurately we follow the directions on the labels.

Almost all of us have one of the most practical disinfectants known to man right in our homes at all times. Ordinary chlorine bleach is one of the most efficient and certainly one of the least expensive disinfectants available. Chlorine bleach will disinfect, deodorize, and remove stains from sinks, bread boards, cutting boards, woodwork, windows, tile, plastic enamel surfaces, etc.

However, chlorine bleach should never be used on steel, aluminum, silver, or chipped enamel because it can cause discoloration and eventually may even cause corrosion. If a chlorine bleach solution comes in contact with metal, it should be rinsed off thoroughly with clear water.

Although directions must always be followed exactly for each brand of chlorine bleach, a general rule of thumb can be followed with all of them: three-fourths cup of chlorine bleach should be added to a gallon of warm water. The surfaces we want to clean should be washed first with soap and water and rinsed. Then the chlorine bleach solution should be applied and allowed to stand for several minutes, then rinsed and air dried or buffed.

A more diluted solution (½ oz. per one gallon water) may be used to disinfect certain items and surfaces by wiping and/or dipping. Surfaces treated in this manner need no final rinsing.

Besides killing, slowing down, or preventing the growth of microorganisms, there are four ways in which cleansers remove soil:

(1) By wetting or reducing the surface tension of water so it is wetter.
(2) By dispersing or breaking up soil into smaller bits and dispersing it throughout the wash water.
(3) By emulsifying (breaking up grease into removable particles).
(4) By sequestering (inactivating the minerals found in hard water so that they do not stand in the way of the cleansing process).

In order to use cleansers safely and effectively, these basic rules should be followed:

(1) Carefully follow directions on the label.
(2) Keep the cleanser in contact with the surface to be cleaned long enough for it to do its job.
(3) Apply enough pressure and scrubbing action to complete the job.
(4) Use only clean water.
(5) Rinse when necessary. (This is particularly important when full-strength cleansers are used.)

LARGE APPLIANCES

Major kitchen appliances are not only work savers, they are also designed to help protect us against foodborne diseases through temperature control and elimination or inactivation of microorganisms.

How many of us realize that our dishwashers, ovens, refrigerators, and freezers also serve as a means of controlling microorganisms? We are told instead how we can plan ahead and save time and energy by using our appliances efficiently. We are informed again and again how we can gain status by making an elaborate dish using the oven or freezer.

Certainly these are bonuses, but when we start to look at our appliances not only as status symbols and work savers, but also as health savers, we begin to recognize the importance of keeping them free of dangerous microorganisms.

A major appliance can be made from any number of different materials which need to be protected during the cleaning process. Clean them often with warm water and a detergent so that dirt, soil, and grease cannot build up, thus requiring a harsher cleanser. A good rinsing followed by a buffing will usually complete the job.

Stoves

If we will use medium-to-low heat when we cook on the burners and when we bake in the oven, food will not boil over as quickly and will not splash or splatter so easily in the oven. This, of course, means the cleanup jobs will be less severe.

When we do spill food on the stove anywhere, we should wipe it up immediately so it won't leave a stain or bake on. When the stove cools off, it should be washed with a sudsy sponge or cloth. Stove tops and ovens should be washed and rinsed after each use. This 2-minute job may mean we will never have to resort to harsh cleansers and a time-consuming cleaning operation.

When the stove is washed, it should be cool and all trays and shelves should be put to soak in a sinkful of hot sudsy water. When all the parts are out, the stove should be washed all over. If the oven or broiler is very greasy, 3 tablespoons of ammonia to a bowl of hot, sudsy water will help dissolve the cooked-on grease. The ammonia solution should be put into a cool oven, the door closed and left overnight. It will loosen the grease enough so it can be wiped up the next morning with a cloth dipped into hot, sudsy water. (Ammonia should *never* be used on aluminum drip pans.)

Spray cleaners or paint-on chemicals should be used with utmost caution, because they are caustic and will burn or eat away other surfaces, including our own skins.

Suds on a pipe cleaner or a percolator brush are effective for washing behind control knobs. When the stove is clean, it should be rinsed all over with a cloth wrung out of hot, clean water, then buffed to shine. Burners should be scrubbed with a stiff brush to remove any food particles, then rinsed and thoroughly dried before they are replaced and used.

The pilot light should be turned off before the burners on a gas stove are removed. The burner holes should be dry before the pilot or the burners are relit.

Electric burner units should not be washed, because spilled food burns off. Burner rims and reflectors or drip pans should be washed in hot, sudsy water, rinsed, wiped, and replaced. (They can be soaked in a solution of 1 cup laundry detergent to a gallon of hot water.) A nylon net scrubber will not scratch the porcelain surface and will remove most baked-on food if it has not been allowed to accumulate for too long. If there is a deep well cooker, the liner should be taken out and washed in hot, sudsy water, rinsed, and wiped dry.

If the surface of the stove is steel or chrome, it should be toweled dry so it won't be marred with water spots.

A long-handled brush can be used to reach under the drip pan area. Harsh cleansers should never be used on any chrome trim, because chrome is a soft metal and will scratch very easily. A polish made just for chromium or other soft metals will usually work on difficult spots.

Teflon-finished, mirror-like metal, or glass parts should never be scrubbed with a harsh cleanser. They should be washed gently with hot water and detergent. The most sensitive portion of any oven is the sensor (or metal tubing) which is located at the roof of the oven in gas ranges and near the back of an electric oven. The sensor must be treated gently and wiped off carefully without jarring, because it controls the temperature of the oven and is very sensitive. Harsh cleansers should never be sprayed or painted on the sensor, and if it accidentally happens, they should be removed gently and quickly.

In an effort to cut down on cleaning, many of us often cover a broiler pan with foil, but this can cause disastrous grease fires if we do not cut slits in the foil to allow the grease to run down through the grid into the broiler pan.

The easiest way to clean a broiler pan is to remove it from the oven while it is still hot and pour the drippings off immediately. With the grid in place, detergent should be sprinkled liberally over both, and the pan should be filled with hot water. Then wet paper towels should be laid over the grid. The steaming will make both grid and pan a lot easier to clean. A steel wool soap pad is useful on baked-on spots, and, of course, an oven cleaner can be used on porcelain; however, it may pit or change the color of the metal.

Foil should never be used in the bottom of ovens or in the drip pans. Doing this can result in poor cooking results as well as possibly burning out the heating elements. Preformed drip pan protectors can be bought in supermarkets and these should be used instead.

Although more and more ranges have self-cleaning or continuous-cleaning features, these only affect the ovens. Routine cleaning is still necessary for the rest of the stove.

Self-cleaning ovens are of two basic types. The prolytic type utilizes a 750° F. temperature, and within 2 to 4 hours reduces all grease and spillovers to a fine ash. The continuous cleaning oven chemically dissipates grease and most spills at oven temperatures ranging from 350° to 500° F. The cleaning takes place when the oven is used for cooking.

The new, smooth-top ranges are far easier to take care of but they still need to be washed often.

Microwave ovens do not heat the oven walls or interior, so when electronic cooking is used by itself, splatters do not burn on. We should remove them immediately with a wet, soapy cloth. Surfaces should then be rinsed and dried.

In order to help keep microwave ovens operating safely and efficiently, they should be cleaned frequently. Scouring pads or steel wool should not be used, nor should any other harsh abrasives or cleaners.

The door will seal far better if the oven surfaces in the interior are kept spotless. Containers should never be set on oven doors because they might spring or warp the door. The only way to determine if an oven door is emitting microwave radiation is to have it tested with a properly designed instrument. Microwave ovens should never be operated when they are empty. We should follow operating instructions carefully.

Refrigerators

Since we know that low temperatures inhibit the growth of microorganisms, we begin to realize how useful and essential the refrigerator should be in keeping down the incidence of foodborne diseases.

A clean refrigerator not only does not harbor molds or other food spoilage agents, but it also stays in good condition longer. However, even low temperatures will not protect us if we neglect foods for an unreasonable length of time and let them spoil. If we neglect our refrigerators, they can become a haven for microorganisms which grow at low temperatures. Odors are difficult to remove from refrigerators. To keep odors from becoming a major problem, we should cover foods securely, rotate them, and clean our refrigerators often.

The frost-free refrigerator, as great a boon as it is, has probably been partially responsible for the development of careless habits. Since we may not be forced to defrost each week or even each year (if we are lucky enough to own a frost-free refrigerator), we tend to forget about regular cleaning.

When food is spilled in the refrigerator, we should wipe it up immediately to prevent stains and microbial growth. Use a clean cloth wrung out from clean, sudsy water and rinse with clear water.

Some bacteria thrive at temperatures of 40° F. (the temperature of the average home refrigerator). So when we leave food spills on refrigerator surfaces, we are asking for trouble.

Our refrigerators should be checked once a week to make sure that nothing has spoiled. Anything that is questionable should be discarded. The inside of the refrigerator should be cleaned once a month. (It is always best to turn off all refrigerator and freezer controls while we are cleaning them. If there are no manual controls, we should unplug the appliances.)

All the food should be removed and all containers wiped down with a clean cloth wrung out from sudsy water.

The inside surface of the refrigerator should be washed down with a solution made from detergent, 1 to 2 tablespoons baking soda, and a quart of warm water.

All shelves, storage drawers, egg storage bins or racks, cheese and butter racks or keepers should all be scrubbed, rinsed, and dried.

We must also give special care and attention to the gasket running around the outside edge of the door. If it is kept clean, it will fit tightly when the door is closed. But if it is covered with grease, dirt, and food spills, the door will not be sealed properly, and the refrigerator cannot stay as cool as it should. Neglect of the gaskets will cause them to deteriorate. Strong detergents are harmful to gaskets, therefore they should be washed with a warm water and soda solution, then rinsed with clear water and dried.

The area where the motor/compressor machinery is located behind the grill at the bottom of the refrigerator should be vacuumed regularly. The condenser coils at the back of the refrigerator should be dusted or vacuumed at least once a year. The evaporator pan where the defrost water collects should be removed and washed. (This can be a haven for the breeding and growth of microorganisms and pests.)

In a self-defrosting freezer, cleaning should be done twice a year or whenever we spill some food.

Frost removal is a major consideration in the care of all freezers. A big frost buildup cuts down on freezer efficiency.

In conventional refrigerators the frost accumulates on the cooling unit. We have to remove the frost by turning off the appliance and allowing the frost to melt. This is not only time-consuming and inconvenient, but the food also warms up during the time the refrigerator is being defrosted. In many combination refrigerator-freezers a special defrost system cycles on automatically to remove accumulated frost. The system is on only for a short period of time once or twice a day, so very little frost accumulates. The water that does collect drips into an evaporator pan and gradually vaporizes into the air.

In frostless models no visible frost forms on the walls, food packages, or ice trays in either the refrigerator or the freezer section, because moisture is removed from the cooled air before it circulates through the food storage areas.

We can hasten thawing in a standard freezer with a hair dryer or pans of hot water set right into the cavity of the appliance. The last thing we should do is try to chip off the frost, because we are likely to puncture the interior surfaces or coils.

After all the frost has melted, the water should be drained out or mopped up, and the freezer should be washed down with the same sort of solution we used to clean the interior of our refrigerators. All racks and drawers should be cleaned this way, and then all the surfaces should be rinsed and wiped dry.

While we are defrosting and cleaning our freezers, we should take care to store the frozen foods by wrapping several items together in layers of newspapers or by placing them in boxes lined with newspapers and covered with blankets. Insulated coolers can also be used. When the freezer is dry and clean, food packages should be wiped dry before putting them back into the freezer.

Moisture is removed when air cools. Outdoors, we know this process results in the forms of snow, frost, and rain. In the refrigerator or freezer, it appears in the form of frost.

Once it was thought that hot food should never be put into the refrigerator, because large amounts would bring up the internal temperature to dangerous levels. However, modern refrigerators can easily accommodate normal amounts of hot foods, and immediate refrigeration greatly cuts down on bacterial growth.

Refrigerator interiors and parts are made of polystyrene, plastics, and metal. Polystyrene may be damaged by bleaches, steel wool, or cleaning powders, but it is easily cleaned with a soda and/or detergent solution.

An all-porcelain interior is acid and stain resistant, but painted aluminum, steel, and plastic can be pitted or stained by spilled milk, acid juices, or even medicinal preparations and drugs.

Plastic and glass may crack when they are subjected to sudden temperature changes, so we should always use lukewarm water with the detergent and soda solutions. Highly perfumed detergents should not be used to wash refrigerators and freezers.

When we store cans, jars, and bottles in our refrigerators, we should first wipe them off with a damp, clean cloth wrung out from a hot, sudsy solution. We should use a clean dry towel to wipe off fresh vegetables before putting them in the refrigerator, and all wilted outside leaves and stems should be removed so as few spoilage organisms as possible will go into the refrigerator. Some vegetables should be stored in airtight containers. Some should be rinsed before storing. (See Chapter 8.)

CLEAN AND HEALTHY KITCHENS

Dishwashers

The average household uses nearly 50,000 dishes, glasses, pots and pans that have to be washed, dried, and put away each year. It is no small task to make sure these are as clean and free from bacteria as possible. (And, of course, under no circumstances should we ever use cracked glasses or dishes. They should be discarded.)

A study conducted by the University of Louisville School of Medicine found that the average bacterial count for all machine-washed dishes was less than one per plate. The bacterial count for dishes washed by hand was as high as 16,000!

All of the automatically washed dishes easily met the Public Health Standard of 100 bacteria or less per plate, a level recognized by health authorities as safe and attainable (11).

Dishwashers are one of the greatest boons to food sanitation and cleanliness . . . if we use them properly. The automatic dishwasher performs by using high temperature, water force, mechanical action, and the chemical properties of detergents. A detergent/hot-water combination is gushed on every side of the dishes, pots and pans. Then the hot water removes the dirt and soil and flushes them away.

Cleaning in a dishwasher is accomplished by a fairly small volume of water. Dishwashers do not fill as completely as clothes washers do. Instead, a dishwasher uses several small fills during a cycle to accomplish the washing and rinsing actions.

The total volume of water used in a complete cycle can vary from 6 to 19 gallons, depending on the number of washes and rinses included in a particular cycle.

Sometimes the water pressure in our homes is reduced because of lawn sprinkling, bathing, and other activities. As a result, water of insufficient amount and temperature may not completely clean our dishes. Therefore, we need to give some thought to exactly when we run our dishwashers.

Another of the more common mistakes we make when we use our dishwashers is to fill the detergent cup with the cleanser a long time before we run the appliance. Unfortunately, the cleaning compound may lose some of its effectiveness if the detergent is exposed to air for a lengthy period before it is to be used. (For this reason, it is important to select only those packages of dishwashing detergent in the supermarkets which are in good condition. Any holes in the package may weaken the strength of the detergent.)

Sometimes we crowd our dishwashers with large pans and dishes which keep the scouring action of the hot water from coming in contact with all surfaces. Frequently, silverware is crowded so badly or nested together that the water and detergent cannot flush out the food particles.

Only detergent which is manufactured specifically for dishwashers should be used. Anything else could result in expensive service calls and repairs.

In order to dissolve detergent and food, dry dishes effectively, and kill bacteria, the water temperature should be between 140° and 160° F.

In hand washing, the hottest water most of us can stand is somewhere between 110° and 120° F.

If our hot water heaters are located a long way from our dishwashers, it may be necessary to raise the temperature setting. Water spotting and incomplete cleaning may be caused by using water below the recommended temperature level. It's a good idea to hold a candy thermometer or meat thermometer under the hot water for a moment now and then to determine exactly how hot the water going into our dishwashers really is.

Most dishwashers are self-cleaning; however, after long usage in hard-water areas, they may develop a white film. The tub should be wiped with a damp cloth and a mild cleansing powder. If filming is severe, we can place two small bowls, each filled with 1 cup of white vinegar in the lower rack, and then run the dishwasher through the regular washing cycle. This vinegar rinse cannot be used more than twice a month, however, because more frequent use can be harmful to the interior surfaces.

Food Disposers

Relatively little care is required to keep our waste or food disposal units clean and odor free. The appliances should be run long enough during each food grinding operation so that all waste food is cleaned out. The drain should be flushed once a week by filling the sinks with two to three inches of water and then releasing the stopper.

Drain cleaning chemicals should never be used in a disposal. If odors do develop, citrus rinds can be ground up to deodorize the units.

The most important advice we can follow is to grind all waste immediately. Foods decay rapidly, and the acids that develop cannot only cause odors, but also corrode the metals in our disposals.

The water flow rate should be fast enough to keep wastes moving down the drainline, enough to fill a quart container in 10 seconds. The disposal and cold water should be left on for a short while after grinding to completely flush out the drain.

Cold water must be used to keep fats solidified and to prevent their solidifying farther down the drain and clogging it.

Disposers are self-cleaning and should require no special care.

Waste Compactor

Routine cleaning of this handy appliance should become part of our regular cleaning plan. Spills should be wiped up as they occur, and the exterior and drawer interior should be washed regularly with sudsy water.

Although some manufacturers say food can be compacted, we should realize that odors develop with the natural decaying processes. Disinfectant sprays cannot stop decomposition. So wastes that must remain in the disposer for several days will probably still develop odors even though they have been sprayed.

When we are ready to compact, it is a good idea to place paper towels or newspapers over food wastes to keep the ram clean.

Portable Appliances

We Americans have had a love affair with portable appliances for thirty years. Because we take them for granted, we are likely to forget they can be dangerous "germ" holders.

Regular cleaning should be done with hot, sudsy water, a thorough rinsing with clear water, and then a good drying. (But no cleaning should be started until the appliance is unplugged.) For stubborn stains, we can ordinarily use steel wool pads. For aluminum interiors of cooking utensils, boil a solution of 2 tablespoons cream of tartar to 2 quarts of warm water in them for 20 minutes.

If we had to decide which three spots in our kitchens spread the most bacteria, we would have to choose the can opener, the meat grinder, and the bread board. Unfortunately, most people overlook all three when they clean a kitchen.

(Actually, the Food and Drug Administration recommends using a separate cutting board for preparing meat and another one for preparing vegetables, cheese, and other foods.)

All appliances that are used for grinding, slicing, and mixing should be disinfected regularly, because frequently pieces of food cling to them after use. This, of course, provides a perfect breeding ground for bacteria.

When we wash meat grinders, we should use a brush to remove particles of food from the inner surfaces. Some experts recommend disinfecting meat grinders with the diluted chlorine solution both after washing and before using again.

If we are grinding a lot of meat, poultry, or fish, we should grind it all at once rather than let the food set out for longer than 30 minutes. That amount of time is just long enough for the microorganisms to grow and reproduce.

A manually operated can opener can simply be dropped into hot, sudsy dishwater and cleaned quickly and easily. An electric can opener poses a different sort of problem. It should be unplugged and rinsed off with a disinfectant at least once a week. Stubborn stains can usually be removed with a non-abrasive cleanser. The cutting wheel should be removed and washed frequently with warm, sudsy water, rinsed, and dried. Of course, the lids of cans should be wiped off with a sudsy cloth before the can opener is ever used.

Bread boards should be washed with hot water and soap and sanitized by dipping them in the diluted chlorine solution, by dipping in a more concentrated chlorine solution and rinsing, or by washing in the dishwasher.

CLEAN AND HEALTHY KITCHENS

Electric mixers should be kept covered so microorganisms and dust from the air do not settle on the beaters and in the bowls.

Blenders should be cleaned by filling the container half full of warm water and adding a few drops of liquid detergent, and then covering and blending briefly. The blender should be rinsed and dried thoroughly, or even washed in the dishwasher if directions permit it. The motor base should be polished with a clean, damp cloth, then dried.

Electric frypans should be cleaned according to manufacturers' directions. We should never immerse a hot frypan in water, because it may cause warping. Some models may be washed in the dishwasher.

We should always wash both the inside and underside of frypans after each use with hot, sudsy water to prevent a buildup of grease deposits.

Toasters should be cleaned with a damp cloth, dried thoroughly, and rubbed with a soft dry cloth after each use. Once a week the toaster should be unplugged from the electrical outlet and held over the kitchen sink. The crumbs should be shaken out and the toaster wiped with hot, sudsy water, then rinsed and dried.

Electric knives are a potential source of bacteria, because the handle cannot be immersed. After each use, it should be thoroughly wiped with a sudsy cloth, then with a clean, wet cloth, and finally polished with a dry cloth.

Ice crushers should be washed in warm, sudsy water, then rinsed and dried.

Coffeemakers should be cleaned after each use. The body of a coffeemaker should be half-filled with water and detergent and washed, then rinsed thoroughly and dried. No electric coffeemaker should ever be immersed in water unless the directions say it can be done safely. Coffee stain remover can be used in a stainless steel pot. It should never be used in an aluminum pot because the cleaner may darken or even pit the interior. A coffeemaker should be stored with the lid ajar to avoid a musty odor.

The interior of toaster-ovens should be washed frequently with warm, sudsy water, then rinsed and wiped dry. Crumb trays should be emptied every few days.

Dishwashing

If we are not fortunate enough to own an automatic dishwasher, or the electrical power has been interrupted, dishwashing by hand becomes a very important part of routine kitchen sanitation. If there is a dangerous communicable disease in the family, we can disinfect dishes and glassware by washing them thoroughly in hot, sudsy water and then soaking them 5 minutes in a solution of 1 teaspoon chlorine bleach to 1 gallon of hot water. Scalding with hot water is also effective. Scalding should be a normal part of hand dishwashing.

The kitchen should be arranged so there is plenty of room for scraping, rinsing, stacking, washing, and scalding.

These are the basic points we need to remember:

(1) We should scrape and rinse heavily soiled dishes.
(2) We should start with the dishes that are soiled the least and work up to those with the heaviest soil.
(3) We should keep the washing solution fresh and sudsy.
(4) Our dishes should be rinsed (scalded) with the hottest water available.
(5) The dishes should be air dried or wiped with a freshly laundered towel.

Dishwashing detergents that are used for hand washing should provide long-lasting suds which dissolve the grease well and hold the soil in suspension easily. There should not be any water spots left on dishes, pans and pots, and it should work equally well in either hard or soft water. Dishes and utensils should have a non-slippery finish.

After dishwashing, the sink should be washed completely with hot suds and rinsed. A dishcloth can be disinfected by soaking in a solution of chlorine bleach and water for a few moments.

Surfaces

Cupboards, drawers, and counter tops should be washed at regular intervals with a disinfectant. Shelves of cupboards and drawers should be lined with paper and changed frequently.

One of the best cleaning solutions for painted walls and woodwork is one we can make for ourselves in our own kitchens for very little cost. It consists of a gallon of hot water, 1 cup ammonia, ½ cup vinegar, and ¼ cup baking soda. Walls should be washed from the bottom up so water running down over dirty walls will not leave streaks. The cleaning solution should be changed when it becomes dirty. Surfaces should be rinsed with clear water.

Of course, many commercial cleaning preparations are very satisfactory and effective. Manufacturers' directions should always be followed carefully, and all cleansers should be kept out of the reach of children or irresponsible persons.

But we cannot keep our kitchens clean if we use dirty tools. Dirty brooms, mops, brushes, dust cloths, sponges, rubber gloves, and carpet sweepers will merely spread dust, dirt, and microorganisms instead of getting rid of them.

Before we shake a mop or a brush, we should cover it with a paper bag, if we cannot shake it outdoors. Brooms should be dipped up and down in hot, sudsy water until they look clean, then rinsed under hot running water. Brooms should be hung, head down, until dry. Mop heads should be removed from handles and washed in the washing machine with rugs.

Nylon bristle brushes were proved to be more sanitary than cloth mops in a study done by medical students at the University of Colorado Medical Center in Denver. In an experiment, seven hospital floors were washed twice—once with a machine with nylon bristles and once by hand with a mop with a cloth head. The bacteria colonies were counted both before and after mopping. The nylon bristles proved superior for six out of the seven times. Bacteria count dropped drastically whenever the nylon bristles were used. The same disinfectant was used under both conditions.

Personal Hygiene

A study made at Cornell University involved pathogens (dangerous bacteria) which most of us have on our hands a good deal of the time. The hands of some food industry workers were covered with fluorescent powder. Investigators were appalled when a black light used to trace the places the workers' fingers had

KEEPING FOOD SAFE

touched, showed crisscrossing of fingerprints from toilet to kitchen to serving utensils to food (12).

Cleanliness in the kitchen can help us to live healthy, active, and satisfying lives. These basic rules can help:

(1) If we are sick, we should stay out of the kitchen.
(2) If we have pimples, boils, or infected cuts on our hands, we should stay out of the kitchen or use disposable, sterile plastic gloves. The same organisms which can cause these problems can also cause food illnesses.
(3) We should wear only clean clothing when we work in the kitchen.
(4) We must always wash our hands thoroughly with hot water and soap after using the toilet or after helping someone else use it.
(5) We must always wash our hands with soap and hot water after smoking or blowing our noses.
(6) Our hands must be kept away from our mouths, noses, hair, and bodies when we are preparing food.
(7) We must cover our sneezes and our coughs with clean handkerchiefs or facial tissues.
(8) We must always wash our hands with hot water and soap after touching any raw meat, poultry, and eggs. We must wash our hands before we touch anything else.
(9) We must never use the same spoon for tasting that we use for cooking and serving. Baby food should never be served out of a jar. Using the same spoon to take out bite-sized servings can contaminate the entire jar. (Not applicable to jars used at one feeding.)
(10) We must wash our hands with soap and hot water after petting animals or touching pets' dishes or bedding.

It is also important to realize that we should not use silverware which has fallen to the floor without first washing it, or that we must not touch the tines of forks, the blades of knives, or the bowls of spoons when we set the table. We should not touch the rims of glassware, bowls, or cups, and we should keep our hands off the serving surfaces of plates.

Our hands can be contaminated dozens of ways, from touching raw meat to sneezing.

And although we have all been washing our hands ever since we can remember, not many of us really know exactly how to go about it.

The important thing to remember is that soap should not always be considered a "germ" killer. It may or may not contain a germicide. Soap's primary function in hand washing is to lift the soil and bacteria and hold them in suspension until they can be rinsed away.

For that reason, we should first wet our hands with warm water, then cover them with soap. We should rub our hands together in a circular motion for 30 seconds, not forgetting the areas between our fingers and under our fingernails. Then we should hold our hands with the fingers pointing down as we rinse them, so the soil and bacteria will drain down our hands, off the fingertips, and into the drain.

And while we cannot (and should not) try to make our kitchens as sterile as an operating room, we still need to clean them regularly in order to keep microorganisms at safe levels. The human body can tolerate bacteria without succumbing to illness. However, each individual differs in his or her tolerance. We must exercise common sense and follow a basic sanitizing routine.

6

We Are Bugged!

In March 1973 the Denver Housing Authority evicted a tenant and transferred his furniture to three different locations throughout the city for storage.

To their consternation, housing authority officials soon discovered the tenant's furniture was infested with cockroaches. What had started out as a routine eviction soon turned into a major eradication crisis.

Not only were the city fathers forced to spend a considerable amount of money to kill the cockroaches, but they also spent several anxious days until the hordes of roaches were eradicated from the storerooms (3).

Although Denver's problem was certainly not unique, the incident does point up the fact that most of us are only a mistake or two away from genuine problems with pests both inside and outside our own homes.

If numbers and varieties mean anything, the sheer volume of insects and rodents in the world is far from reassuring. There are some 600,000 kinds of insects as compared with only 10,000 species of mammals. The actual number may be four or five times this figure (2)!

When we consider bulk, the picture is really ominous. It has been estimated that the combined bulk of all the insects on earth is equal to that of all land animals living today!

When a pest like the rat or the German cockroach invades our homes, we begin to realize how formidable an enemy an insect or rodent can be, with their cunning, their maneuverability, their amazing life processes and reproductivity.

Most of us think of pests as only irritating nuisances. Many of us are ignorant about the ways to prevent and/or eradicate them. How many of us realize pests can cause unbelievable damage, poverty, hunger, illness, and even death?

When Denver officials were figuratively wringing their hands over their cockroach dilemma, they were acting out a modern version of an ancient concern with pests. For insect control was a real problem for the ancients, and eradication was impossible. Through the centuries various beliefs sprang up about the origin of insects. Even in the late 1800s, many people thought insects came from boils in the skin, stale perspiration, and from rotting wood and food.

Even though insects have always been feared and looked upon with disgust, they have always been respected. One of the most famous of all passages in recorded ancient literature tells about the plague of locusts sent by God to help Moses free the Jews.

Interestingly, some of the most unpleasant insects known to man were used in medicinal practices in early cultures. There were preventive and eradication methods, but most of them did no good at all, except to bolster spirits. Charms or amulets were commonly used in Egypt, and sacrificial rituals were practiced in the Orient. In addition, our ancestors had and used large numbers of lotions, minerals, and herbs, but most of them did absolutely no good at all.

However, we now recognize that a few old-time practices were based on sound reasoning. Herodotus, for example, wrote about the learneds in Egyptian temples who shaved their bodies completely every few days and washed frequently to keep themselves free of insects.

In 1737, Dr. Boyle Godfrey of London performed a series of

WE ARE BUGGED!

"Experiments to destroy Buggs in Houses" and the results were duly recorded in *Miscellanea vere Utilia:*

> At last I tryed the following which no animal life can subsist under: viz Let matches of Sticks be made by dipping them in common sulphur so that they have adhering to them about 4 lbs., which will do for some rooms (but others require about 6 lb.) which matches must be set upright in a large Earthen pot . . . and set on fire in the Room where the Buggs are troublesome, stopping the Chimney by hanging Blankets before it and likewise . . . against the cracks of the Door . . . This sulphur, while burning will give a prodigeously strong Funk, and such as will kill all Creatures in the Universe. This might be commenced in the morning, and by next Day the Fume will be subsided. It would be proper to make the Fume when the hot weather comes on and they begin to bite; and again in September.

Fortunately, such heroic measures are no longer necessary to rid our homes of pests. Today, we know that most household pests can be controlled by systematic and effective housecleaning and the use of pesticides. It is a lot easier to prevent pests from infesting an area than it is to get rid of them once they get inside.

There is some experimenting presently being conducted with controlled temperature and atmosphere studies. Insect activity practically stops at 45° F. and below. Insects cannot survive in atmospheres with less than 2 percent oxygen. So in the future, it may be practical to use temperature and atmosphere controls to guard against insect infestation.

Household Pests

There are more pests around than most of us realize, but we are only concerned here with some of those which will contaminate or destroy our food.

There are libraries full of information about their control, but few of us take the time to read the information available. The Federal Government has numerous bulletins on the general subject of pest control. But few of us know how and what to order.

As a result, a trip to the hardware or garden store to buy a pesticide can be a confusing experience. But there are specific measures we should follow for each of the pests which threaten our food supply.

Cockroaches

The cockroach (as the Denver housing authorities know all too well) grows from 12 to 35 mm. (or ½ to 1½ inches) long, with whip-like antennae and two pairs of wings. The front pair are rather stiff and rest over the hind pair when a cockroach is at rest.

House-infesting cockroaches vary from a pale to a reddish brown to an almost black color, and their bodies are smooth and shiny.

They prefer a warm climate, but there are certain species which live in cooler climates and have become rather cosmopolitan and have learned to enjoy the luxury of our centrally heated homes, warehouses, and offices.

The cockroach particularly relishes meals of cereals, almost any sweet substance, and meat products. But they are not averse to feeding on leather, hair, wall paper, and dead animals.

The three essentials which cockroaches must have are warmth, moisture, and food. They hide during the day in sheltered, dark places, and they come out at night to forage for food. They feed on garbage, and because of this, they can transmit many diseases. (Cockroaches deposit waste matter from their intestines wherever they crawl.) Studies have shown that several types of bacteria can emerge from the intestinal tract of a roach, alive and ready to continue growing and reproducing. One experiment in Brussels, Belgium, showed that cockroaches can transmit *Salmonella typhimurium* (2).

Good housekeeping and the efficient use of insecticides are the best measures we can take against cockroaches. If we are ever suspicious that we might be harboring them in our homes, we can wait until dark and then enter the suspected area and switch on the lights. The chances are, we will find several wandering cockroaches which have come out of their hiding places.

Diazinon, malathion, or ronnel will control all kinds of cockroaches. Chlordane or lindane will control all kinds, except

WE ARE BUGGED!

possibly the German cockroach (12). (In some areas, the German cockroach has developed resistance to these materials.)

Either a household surface spray or a dust can be used. For a severe infestation, both can be used. We should apply the spray first, then the dust. The dust should be forced into the cracks and openings which are difficult to reach with the spray.

An ordinary household spray gun is best to use, or one of the surface sprays we can buy in retail stores in pressurized cans. We should apply enough spray to moisten surfaces thoroughly, but not enough to drip or run.

A surface spray or aerosol mist containing pyrethrum can be used to get down deep into cracks, crevices, and other hard-to-reach places. This usually will not kill roaches, but it will drive them into the open where we can kill them with a surface spray or dust (12).

Not only should we apply insecticide to the areas affected, but also to these spots:

(1) Beneath the kitchen sink and drainboard.
(2) In cracks around or underneath cupboards and cabinets, and inside them, especially in upper corners.
(3) Places where pipes or conduits pass along the walls or go through them.
(4) Behind window or door frames.
(5) Behind loose baseboards or molding strips.
(6) On undersides of tables and chairs.
(7) On closet and book shelves.

When we have to treat cupboards and pantries, we should take everything from the shelves and remove drawers so foods and utensils will not become contaminated by the insecticide.

It isn't necessary to treat the inside of drawers if we have thoroughly cleaned them. However, it is important to treat the sides, backs, and bottoms of drawers and the insides of the cabinet.

Residual sprays should be applied with care in our kitchens or pantries to avoid any possible contamination of our food or utensils.

Many insecticides will kill cockroaches and can produce a satisfying number of dead ones, but complete eradication is hard to achieve, because most of the roaches and certainly the egg cases are likely to be hidden away where nothing will reach them but a penetrating fumigant.

For this reason, more cockroaches may well appear after several weeks following an apparently successful treatment.

Common sense tells us that we should seal cracks and crevices as a preventative measure even if we do not have cockroaches. We should be able to open heat ducts from time to time for inspection and treatment if they cannot be completely sealed off. And, of course, we need to eliminate any easy sources of food for cockroaches, like food particles, debris, or rubbish. Cockroaches are usually an indication of poor housekeeping.

Houseflies

An average fly is about 6 to 7 mm. (or nearly ⅓ inch) long. Houseflies breed in decaying organic matter such as the excrement of various animals, decaying animal remains, rotting vegetable matter and garbage, as well as the food on our tables.

The essential needs of houseflies are carbohydrates (mainly sugars), protein, and water. But sugar is the most critical need, and flies will die more quickly when it is taken away than if water is eliminated from their environment.

Flies contaminate everything they touch, and they spread many, many human diseases. The housefly feeds on fecal material, vomit, sputum, and then is likely to land on our food.

The physical makeup of the fly is particularly well adapted for the transmission of disease organisms because its body is covered with fine hairs. Furthermore, the fly has the particularly nasty habit of excreting and regurgitating everywhere it comes to rest.

The control of fly populations has been shown to reduce enteric infections such as shigellosis, and one study in 1949 showed that fly control reduced infection, disease, and death due to diarrhea (15 and 2).

Other studies have shown that houseflies transmit *Salmonella* (4).

Mature houseflies can live as long as 12 weeks, but generally their span of life lasts no more than a month.

WE ARE BUGGED!

The total number of flies in a neighborhood indicates how well the residents maintain outdoor hygienic measures. Since livestock in urban areas is no longer of any significance, flies are attracted to our domestic refuse. If this is thoroughly and properly taken care of, flies are not particularly bothersome.

The general rules for controlling houseflies are simple:

  (1) Sanitation measures must be observed and practiced.
  (2) Screens must be used on windows and doors.
  (3) Insecticides must be used properly when needed.

The surest control of flies lies in prevention of breeding in unsanitary areas. Garbage cans should have tight-fitting lids, and garbage must be disposed of at least once a week and more often in summer, if possible.

Garden compost should be covered with mesh screen. Each time a layer of organic waste such as garbage is added to a garden compost pile, it should be covered with a layer of damp soil.

Pet droppings should be disposed of promptly and buried, if possible. We should never allow pet food to stand where it will attract flies.

If there are only a few flies in our house, we can probably get rid of them more efficiently with a fly swatter, but if they are a major problem, we should apply a household spray or aerosol spray. We must be sure the label specifies the spray is for flying insects. And, of course, we should use insecticides only when needed and then handle them with great care. We should follow the directions and heed all the precautions on the labels.

Ants

A picnic on a tablecloth spread on the grass is the fastest and surest way to attract ants. But we can also unwittingly attract them into our kitchens. A stray crumb, an unnoticed crust on the kitchen floor, and we may have a problem. Several different species of ants, different in color and size, like to invade our homes.

Sometimes we may confuse ants and termites. Yet the two insects can easily be distinguished. Ants are constricted or "pinched-in" at the waistline. (See Table 5.) Termites have no constriction at the waistline. Then, too, the rear wings of an ant are considerably smaller than the front wings. There is little difference in size between the rear and front wings of a termite.

As a rule, ants are either blackish, brownish, yellowish, or reddish in color, although sometimes they can be a combination of these colors. Some tropical ants have unusual metallic hues.

Most of the ants we see in and around our homes are natives, but the Argentine ant has replaced many native species in some of our coastal areas. The Argentine ant tends to be prolific and is likely to replace the natives; thus, it has become a major pest in some areas. In bad weather, Argentine ants are likely to group and march into a house and settle down near the furnace to keep warm.

The thief ant and the red Pharaoh ant are others which like to share our dwellings. As a rule, we can find their nests merely by watching the columns as they come and go.

In some areas of the United States the Pharaoh ant is becoming a major threat to hospitals, because they contaminate sterile areas. The ants often crawl over toilet seats, down into sewers and in other unsanitary places.

Ants crawl over any food they can reach, spoil it for human consumption, and carry bits of it back to their nests. As they crawl over clean food surfaces, they spread various kinds of organisms. Many of these can be pathogenic organisms.

They generally cannot attack fabrics, leather, or similar materials. They seldom attack perfectly sound wood, but some species damage wood structures, particularly old houses, by building their nests in the decaying woodwork (12).

The only sure method of completely ridding a house of ants is to find and destroy the nest. If it is outdoors, it should be dug open and filled with boiling water or with two cups of carbon tetrachloride. If benzene hexachloride is used, the nest should be covered afterwards with damp soil to keep the fumes inside. Of course, several insecticides are also very effective.

One way to destroy nests of garden ants is to simply sprinkle them with a 0.2 percent emulsion of chlordane from a watering can. The solution at this percentage is too weak to harm the plant life in the garden.

The most effective way to get Pharaoh ants out of buildings and homes is to treat the area surrounding the nests with an insecticide with a residual nature.

Inside the house, ant nests may be within a wall or partition, under flooring, under a pile of papers, or in an out-of-the-way corner. Then liquid household insecticide containing chlordane, diazinon, lindane, or malathion should be used.

Insecticide should be applied to surfaces over which the ants are crawling in their line of march.

In addition, all cracks, openings, or runways inside the house should be sealed and treated. Also, we must treat the following areas:

(1) The lower part of window frames and around doors.
(2) Supports, posts, pillars, or pipes that the ants might use as runways into the house.
(3) Cracks in baseboards, walls, floors and around sinks, bathtubs, toilets, and kitchen cupboards.
(4) Openings around electrical outlets and plumbing or heating pipes.

If the ants continue to appear after a few days, they are probably entering over surfaces we have not yet treated, so we must search and treat further.

Ants can be prevented from crawling up the legs of furniture by smearing the legs once or twice a week with a strong insecticide or insect repellent.

Mice

At one time or another most of us will find a mouse in our homes. And mice can be a source of a great deal of annoyance and damage. The little rodents usually come into houses from the outdoors when the weather turns cold in the fall. They will slip quickly into a house through a door held open too long or through a large crack. Sometimes they will migrate into our garages or basements through an open window or door, then slip into our houses unnoticed later on.

Interestingly, the word "mouse" has been traced to the Sanskrit word "musha" which is derived from a word meaning "to steal." The mouse is believed to have spread into Europe from Central Asia. It was a simple matter for the mouse to migrate via ship to America.

House mice have learned to enjoy the same foods we like, such as meat products, grains, cereals, seeds, fruits, and vegetables. They eat and contaminate our food, injure fabrics, wood, and other materials and can transmit several human diseases.

Mice spread diseases to humans by biting them, by infecting human food with droppings or urine, indirectly through the cat or the dog, through blood-sucking insects or mites or by dying in a water supply and contaminating it.

Their droppings may infect foods with *Salmonella* bacteria. Plague and epidemic or murine typhus are carried by infected fleas on mice.

The first steps in controlling mice are to seal any holes in our walls, floors, and foundations, and to see that food is not left in places where mice can get to it.

If we have only two or three mice in our house, we can usually kill them with ordinary snap traps, which should be placed along the walls and near holes. They should be placed at a right angle to the wall so the trigger mechanism will intercept the mouse's probable route of travel.

One of the best baits is made by smearing peanut butter over the trigger surface of a trap. Other good baits are cake, flour, bacon, nutmeats, cheese, and soft candies, particularly milk chocolate and gumdrops.

Where mice are so numerous that trapping is impractical, poison bait can be used. Only materials labeled for this purpose should be used. Directions must be followed carefully.

Poisoned bait must never be placed where there is any danger at all of contaminating food supplies. No pesticide should ever be left within reach of children, irresponsible persons, pets, or livestock.

But a qualified pest control operator (exterminator) may be needed to get completely rid of mice.

WE ARE BUGGED!

Rats

The most repulsive of all household pests is the rat. Yet the rat is one of the most adaptable and cunning of all animals. It has not only learned to adapt itself to almost any environment, but even to thrive under the worst of conditions.

Some of mankind's most severe problems throughout the ages can be blamed either directly or indirectly on the rat. London's great plague and the "black plague" which killed more than 25 million people in the fourteenth century were caused, at least partially, by vast numbers of rats.

The source of plague, of course, is a plague-infected rat which harbors a plague-infected flea, which in turn bites and infects man.

Today epidemics of the scale of the Black Death are no longer experienced in Europe, but it has been estimated that from 1898 to 1923 more than 11 million lives were lost from plague in India, China, Mongolia, and other parts of Asia.

Plague outbreaks are reported in this country every few years. However, they are never widespread and are stamped out before the disease is allowed to spread.

The disease is a bacterial affliction of the circulatory and respiratory systems. Humans can be affected by plague in four ways:

(1) Bubonic plague: The blood becomes infected and the bacilli are taken up in the glands. This is the most common form of plague, and it is caused by flea bites. However, bubonic plague can also be caused by infected dust and dirt coming in contact with open cuts and sores.

(2) Septicaemic plague: This was the great Black Death plague. The glands do not control the bacilli, and many hemorrhages occur under the skin, causing it to turn black. It is also caused by the bite of an infected flea.

(3) Pneumonic plague: The bacilli collect in the lungs. This type of plague is spread easily through coughing and from eating contaminated food.

(4) Sylvatic plague: This is now found primarily in ground squirrels, wood rats, deer mice, and woodchucks. Cases

and a few deaths have occurred in the past few years. This is often caused by children's handling of dead, wild rodents.

The plague is bad enough, but that is not the only disease rats spread. Murine typhus fever, infectious jaundice, rat-bite fever, trichinosis, Rickettsial pox, poliomyelitis, and even rabies are suspected by the U. S. Public Health Service of being spread or at least influenced by the presence or activity of rats.

According to an article on rats in the U. S. Public Health Service *Rodent Control Manual* in 1949, the Typhus Control Unit determined that both rats and mice suffer from intestinal infections. These infections are spread to man when he eats infected foods which are contaminated by the excreta of infected rodents. Acute food poisonings of this type are probably much more common than generally realized, and may involve a large number of persons at one time.

Literature through the ages has been filled with references to the rat and the miseries it has caused mankind. A work written in the second century A.D. refers to rats which migrated across rivers, forming living rafts, each rat holding by its teeth to the tail of the rat in front.

The first black rat probably reached the United States in the first ships that sailed from European ports, and by 1927 there were black rats reported in every one of the United States.

Rats will eat anything if forced into it by starvation. They will destroy and most certainly pollute human food, transmit diseases, and damage property. If they are cornered, they are dangerous and will even attack people or pets. (Even though cats are supposed to be a natural enemy, most will not fight a full-grown rat. Owls are the best rat catchers.)

Rats enter our houses to find both food and shelter. There are four essential measures we must know and follow if we ever have to rid our homes of rats:

(1) They can be driven out by starvation.
(2) We can remove their shelters, thus removing part of the attraction.

WE ARE BUGGED!

(3) We can kill them.
(4) We can keep them out.

The most important signs of rat infestations are droppings, runways, tracks, gnawings, a distinctive rat odor, and, of course, the presence of either live or dead rats.

The first and most effective method of reducing rat populations is to decrease their food and remove their living and nesting places.

We must keep our storage places orderly and clean. Lumber and cartons in our basements and storerooms should be placed on racks at least 1 foot above the floor. A room filled with a tumbled mess is likely to become a literal rat's nest.

If our homes have double walls with spaces between ceilings and the floors below, we need to make sure the spaces are tightly sealed off. Rats make their homes in these areas if they are accessible.

Our garbage should be stored in cans with tight-fitting lids and should be removed at least once a week. Spilled food should be cleaned up immediately, and especially should not be left in crevices or under low shelves.

Food should be stored in tin cans with tight covers or glass jars with screw-on lids. Since rats must have water in order to live, they can be forced from a house by removing their water supply.

All holes in exterior walls should be closed off. We should see that spaces around doors, windows, and other necessary openings are no larger than ¼ inch across.

If rats are a serious problem in our neighborhoods, we should install self-closing devices on frequently used doors which open to the outside.

Where rats are a neighborhood problem, community action should be taken. Assistance should be asked of local boards of health.

Pest control operators (exterminators) are trained in the control of rats, and they have the experience and equipment for rat control.

It is not at all unusual for rats to gnaw their way through buildings with wooden floors and foundations, so these floors and foundations should be replaced with concrete when possible, or the building should at least be enclosed with a barrier of metal, concrete, or brick.

The openings through which pipes pass through wood siding should be covered with 24-gauge, galvanized sheet iron flashing.

Ventilators should be covered with 18-gauge, and low windows can be fitted with similar flashing. Wooden doors should be trimmed on the bottom with sheet iron flashing, and the jambs and sills should be trimmed. There should never be more than ⅜ of an inch opening below any outside door. It only makes good sense to keep all outside doors closed when they are not being used.

It is possible for a rat to wedge itself between a drainpipe and the wall and work its way to an upper story, but a circular rat guard will prevent this. Another common way for rats to enter homes is to drop from tree limbs which hang over roofs. Limbs of this sort should be pruned back.

Poisoned bait is recommended as the best means of killing rats (12). We should buy a bait labeled for this specific purpose, and of course we must always follow the directions and observe the precautions printed on the label. Poisons should never be left within the reach of children, irresponsible persons, pets, or livestock.

If a poisoned rat dies in an area we cannot reach, the stench may be unbearable for some time. Therefore, it may well be wiser to have a professional exterminator handle the problem.

Prebaiting is generally successful because the cunning and instincts of rats make them avoid new objects. So in prebaiting, rats are fed an unpoisoned food. After they have learned to eat and like the food, the same kind is poisoned and the rats will usually eat it and die.

Traps are also a very effective means of killing rats in our homes, but the ordinary person does not have the skill or the time to use them. Traps are recommended only where infestation is small or as a follow-up measure after poisoned bait has been used.

Killing rats with poisoned baits or traps will reduce the number

WE ARE BUGGED!

of rats, but the reduction will be only temporary if sanitation and rat proofing are not practiced and constantly maintained.

We should remember that even if the rats are killed, their fleas and other parasites may still be alive in our homes.

Pantry Pests

Several kinds of insects can infest most of the dry food products we keep on our cupboard and pantry shelves. Although many people refer to all of them as weevils, most are actually beetles or moth larvae.

Pantry pests can be controlled by all five of the following measures:

(1) Keeping pantry shelves clean.
(2) Applying insecticide.
(3) Inspecting food packages.
(4) Sterilizing or destroying doubtful products.
(5) Storing food in closed containers.

Food gets spilled. Particles sift out of packages and stay on the shelves or lodge in cracks and corners. Insects can live on this material and get into food packages. Shelves should be kept clean to avoid later infestation as well.

If we are troubled by persistent infestations of pantry pests, we can treat our pantry shelves and kitchen cupboards with insecticides. But the shelves and areas should be washed first. We should apply a household surface spray containing no more than 2 percent of malathion.

We should spray lightly, never overspraying. We should spray only after we have removed all food and utensils.

Almost all dry packaged foods are subject to infestation. Pantry pests live in spices, especially red pepper, paprika, and chili powder. We should inspect all packages for breaks before buying, and we should never put infested packages on our shelves.

The beetles which are included under the general classification of Pantry Pests are easily recognized. Their main characteristic is a beetle-like appearance with stiffened forewings which cover the back wings when they are at rest. They tend to bore right through various types of packaging and on into the foodstuffs.

Moths which attack stored food products are small and mostly buff or gray-colored. They have cream-colored larvae. Nearly 40 different species have been found in various types of stored food products. Moths can infest flour mills so heavily their webbing can clog up the machinery.

Insect pests which attack and consume stored food are nutrition conscious, because they eat only the better parts.

We generally do not realize the vast amount of food destroyed each year throughout the world by insects. We are merely disgusted and annoyed when they get into our own pantries.

We probably would not become ill if we ate an insect part— we undoubtedly have. (If we knew it we might get sick just from thinking about it.) But the real problem with the presence of insects or their parts is that they indicate a lack of good hygienic standards.

If we suspect we have an infestation of insects in our own stored foodstuffs, we should search through all our products and discard any which are badly infested. Then the entire storage area should be cleaned, well ventilated, and treated with an inoffensive insecticidal spray. If this is not done, infestation will occur again, because many pantry pests can bore right on through paper, cardboard, or thin plastic.

If dry foodstuffs are only mildly infested, we can generally sterilize them in the oven at 140° F. for half an hour or hold them in our home freezers at 0° F. or lower for 3 or 4 days.

Small packages can be heated or frozen as they are. The contents of larger packages can be spread out on cake or pie pans so that heat and freezing temperatures can easily penetrate.

Uninfested, heat-sterilized or freezer-treated foods should be stored in clean metal or glass containers that have tight-fitting lids, such as coffee cans or fruit jars.

Even though some people do refer to all pantry pests as weevils, only a few actually deserve the name. If they were the size of an ordinary house dog, weevils would be fearful creatures, indeed. They have a snout-like prolongation of their heads between their eyes, and carry their mouth parts at the tip. But fortunately the weevil is small and is a vegetarian and is inclined to attack grain and cereal products.

Grain weevils are common in warm climates. They are wingless, so they are better adapted to living in stored grain rather than field-growing grain.

The female weevil deposits an egg in a small hole she has bored in the grain with her long snout, and then she seals it with a secretion. The larvae begin life inside the grain, and this, of course, is why we can bring home a completely sealed container of grain, open it and find weevils inside.

The bean weevil is about 3 mm. (or ⅛ inch) long, and lays its eggs on bean pods which are growing in the field. If stored beans are found to be infested, they can be treated by heating them at 145° F. for 2 hours.

All stages of the bean weevil can also be killed by keeping the beans in cold storage at 32° F. for 58 days.

The weevils we see in rye, buckwheat, cereals, and wheat products are rice weevils. They can also injure apples and pears by sucking the juices. The larvae are white with brownish black heads.

Fruit, Vinegar, or Pomace Flies

These small flies can pass through ordinary screening, and they are common in our homes, in restaurants, fruit markets, canneries, and other places where fruit is handled frequently or stored.

The flies are attracted to human and animal excrement and will also feed on fruits and uncooked food. A form of diarrhea, "vin cochylise," is common among vineyard workers, and it is caused by contamination of the grapes by the fruit fly.

D. melanogaster is the best-known species of fruit flies, and an average specimen is 3 mm. (or ⅛ inch) long. It has a tan-colored head with a blackish abdomen, and is found wherever fruit and other organic materials are allowed to rot and ferment.

Its larvae are small, eyeless, legless maggots, pointed at the head end. They sometimes are found in canned fruits and vegetables which are not sealed well. The maggots can be found near the tops of the jars and live in the brine-like liquid.

The adult flies lay their eggs around the edges of the covers on the jars as well as around the spigot and holes on vinegar and cider barrels. They are also likely to infest rotting bananas, pineapples, tomatoes, pickles, and potatoes. They will also get into wine, beer, cider, and vinegar. The eggs are very difficult to see without a microscope. The larval period lasts from 5 to 6 days. When the newly hatched flies emerge, they are immediately attracted to almost any light source. The adults mate often and with overwhelming success.

When we find fruit flies in our homes, we should search for rotting fruits or vegetables or other organic matter. (All opened jars of fruits or vegetables should be kept in the refrigerator. Fermenting vegetable or fruit juices are the most likely sources of breeding.)

But sometimes when we have a heavy infestation, there may be no rotting materials anywhere in the house, no open jars of fermenting liquids or organic matter. Then the trouble can often be traced to dishwater which leaks from sink drains, water which drains from refrigerators and even from overflow drains in our sinks. Many times these areas harbor food particles and offer perfect breeding conditions for fruit flies.

Another source of breeding is an unclean, damp mop head which can sour and breed thousands of fruit flies, and the same can be true of damp, sour brooms and dirty, damp rags.

Mites

Mites can cause severe tainting of foods if they are numerous, but fortunately we can pretty well eliminate the problem if we avoid storing food for long periods under very humid conditions.

Mites are only 0.5 mm. long (or about the size of the eggs of many other pantry pests). Without a microscope, they look like little white specks which will slowly work their way from a source of light. When mites are very numerous, they look almost like buff-colored dust, and lots of times there will be a distinctive mitey aroma.

Acarus siro is the most troublesome mite in stored food. They are found not only in stored grains, in mills and bakeries, but also in our own homes. They particularly love cereals and cereal products and will tunnel right into the germ of the grain, eating only the best and most nutritious parts.

Tainting will occur if there are more than 500 mites per 100 grams of food, but most of us will lose our appetites if we see even one.

If mitey flour is used in pastry mixes, the cooked product will be almost inedible, because of a very bad taste. If it is used in bread baking, the bread will taste sour, have a bad odor, and will not rise.

The best way to guard against mite infestation is to keep cereal and cereal products very dry, so the total moisture content is below 12 to 13 percent.

Stored food products made of cereals or grain should be kept in well-ventilated areas and never allowed to touch any surface which might retain moisture, such as stone or concrete walls or floors.

And, as with any household pests, the best method of controlling mites is by preventing them from infesting in the first place.

Packages of foodstuffs should be sealed with wax paper, foil, or plastic sheeting, but the wrapper must be absolutely secure. Wrapping, especially plastic, can actually encourage the entry of mites because it retains moisture, so the seal must be perfect.

Occasionally, mites will also attack cheeses which are stored for lengthy periods, but this can be prevented by dipping the entire cheese in wax to provide a complete coating.

Mites can be destroyed by heating and by various fumigants which are used for stored product pests, but the fumigants are really rather impractical to use in a home situation.

Keeping Pests Under Control

The old, common sense measures are still the best, steps like keeping windows and doors tightly screened and making sure that screen doors swing outward.

Household pests need food and water and places to hide and breed. If we rob them of these necessities, pests will begin to look elsewhere for more comfortable accommodations.

The basic rules are simple, but sometimes we grow lax. And then it isn't very long before we begin to notice pests who dropped in, liked what they saw, and decided to stay.

We should:
(1) Promptly dispose of garbage, bits of food, crumbs, scraps of fabrics, lint, and other waste materials that pests may eat or in which they may breed.
(2) Keep all foods in tightly closed containers and keep the containers clean outside as well as inside. Before purchasing dry foods, we should examine the packages carefully for evidence of breaks and insect infestation.
(3) Do not permit insect pests to hitchhike into our homes. Cockroaches, for example, often enter a house in the crevices of cardboard cartons used in moving groceries from the store to the home. These cartons should never be left in our kitchens or our basements where pests might hide and breed in them.
(4) Permanently seal up places where pests might enter. We may not be able to close them all, but we can close most. We should calk the openings and cracks around washbasins, toilet bowls, water pipes, drain pipes, and radiator pipes. We should also see that the cracks around baseboards and between floorboards are filled, and we should cover all openings where rats or mice might enter. Windows or doors must be tight-fitting.
(5) Practice pest-prevention measures all the time. The application of pesticides may be needed to supplement good housekeeping.

Proper Use of Pesticides

We should buy only products specifically labeled for the control of the insects we want to kill. And, of course, we need to make it a regular practice (even a habit) to always read the label on insecticide containers each time before we use them. We must follow all directions and observe all cautions to the letter.

Used improperly, many household insecticides can become deadly to us, our loved ones, and even to pets and livestock, and used improperly, insecticides can cause far more immediate damage to our health and our lives than the insects we might be attempting to control with them.

Oil-base sprays should be handled as though they were flammable.

All pesticides should be kept in closed, well-labeled containers in a dry place. They should never be kept under the sink or where they could contaminate food or animal feed. They should be stored in their original containers and the labels must stay intact and readable.

Containers should always be properly disposed of when they are empty. This means we should see that they are collected by a trash service which will bury them in a sanitary landfill dump or will crush and bury them at least 18 inches deep in a level, isolated place where they will not contaminate water supplies. When we throw them into our garbage cans, we should first wrap them thoroughly with several layers of newspapers. Pesticide containers should never, never be reused.

When a label warns against breathing pesticidal mists or dusts, we should open windows and doors when we apply them.

We should avoid getting pesticides on our skin, and we must keep them out of our noses, eyes, and mouths. If we spill any on our skin and/or clothing, we should remove the contaminated clothing immediately and wash our skin thoroughly with soap and water and launder the clothing before wearing it again.

We should never smoke while handling a pesticide, and our faces and hands should be washed with soap and water each time we use one.

If we accidentally swallow some pesticide, we should read the antidote instructions on the label and contact a physician at once.

The information on the label must be read to the physician, naming the active ingredients. If we cannot immediately reach a doctor by telephone, we can call our fire or police departments. They can help us locate a doctor or get us to a hospital. Some cities have poison control centers at one or more hospitals.

Pesticides and Their Application

There are many different types of pesticides, just as there are many different varieties of pests. Insecticides control insects, miticides control mites, and rodenticides control rodents. But we can apply pesticides in different forms and in different ways to serve our specific purposes.

SURFACE SPRAYS

Surface sprays are made to be applied to areas in our homes where insects are likely to crawl. The spray particles are coarse, and they dampen or wet the surface. The deposit will continue to kill insects which crawl over it (12).

Surface sprays can be bought in either pressurized containers or liquid form. Liquids can be applied with a household hand sprayer that produces a continuous coarse spray.

Oil-base insecticides should be used with caution. For example, they will dissolve asphalt tiles. They may also soften and discolor some linoleums and some of the plastic materials kept in our kitchens. If we are ever in doubt about spraying surfaces of this sort, we can test the spray on a small inconspicuous spot.

When we apply oil-base insecticides to the cracks in a parquet floor, we should use a light touch. An excessive amount can dissolve the underlying cement, and the dissolved cement can stain the floor (12).

No pesticide sprays should ever be used in an area until all food and cooking utensils and dishes are covered and small children, pets, livestock, and aquariums are removed.

SPACE SPRAYS AND AEROSOLS

These sprays are made to be applied directly into the air, and they are especially effective against houseflies and other flying insects. They can also be used to penetrate the hiding places of other insects and drive them out into the open where they can be killed with a surface spray or dust.

The particles or droplets of a space spray are much finer than those of a surface spray, and they float in the air for a short while. And the particles of an aerosol spray are even finer than those of a space spray, so the particles from an aerosol spray will float in the air for an even longer time.

Space sprays leave very little residue, and so we shouldn't use them as surface sprays. Aerosols are also usually too fine for surface application.

Some sprays sold in pressurized containers may be labeled for both surface and space applications. If we use one of these products for spraying in our kitchen and pantry, we should first cover the eating and cooking utensils and remove foodstuffs, pet food and dishes.

Some people may be allergic to the materials in space sprays or aerosols. After application, we should leave the room, close the door, and not re-enter for half an hour or longer. We should breathe as little as possible of the chemicals in the sprays.

DUSTS

Insecticide dusts usually contain the same active ingredients as sprays. They are used for surface applications and can be blown out by a household hand duster into cracks, corners, and other places difficult to reach with sprays.

PAINTBRUSH APPLICATION

Insecticide may be applied to surfaces exactly where we want it in liquid, cream, or paste form with a paintbrush. It is particularly effective where only spot treatments are needed. Cream or paste insecticides are usually available in stores where we buy liquids or dusts.

POISONED BAIT

Poisoned bait, as the name implies, is a bait on which a pest will feed and to which a pesticide has been added. Poisoned baits are used to control rodents and some other pests in our homes. Frequently, however, they are more hazardous to us and our pets than any other forms of pesticides. If we use a poisoned bait, we should handle it with extreme care, follow the directions, and observe all cautions on the container label.

Today protection against pests is greatly enhanced because of our increased knowledge, recognition of the dangers of pests, better sanitation, and scientific measures such as the use of pesticides.

TABLE 5
COMMON HOUSEHOLD PESTS

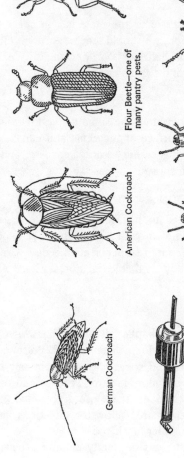

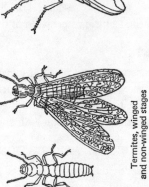

(Reprinted from "Controlling Household Pests," U. S. Department of Agriculture Home and Garden Bulletin No. 96.)

7

Putting Up Food

Late in the eighteenth century, Napoleon offered a 12,000-franc prize to anyone who could come up with a method of preserving food which would solve the spoilage problems his troops faced, and at the same time be superior in taste to the old methods of drying and pickling (5).

After fourteen years of work and study, a French confectioner, Nicholas Appert, won the prize in 1809 by developing a method of preserving food in a bottle. Appert wrote a treatise entitled, "The Book for All Households; Or the Art of Preserving Animal and Vegetable Substances for Many Years" (2).

Appert had no way of knowing why his method worked, yet it was used for years by commercial canners. Later, it was Pasteur who pointed out the relationship between microorganisms and food spoilage.

When we view it against all of mankind's efforts throughout the ages to preserve food, the process of canning is relatively new. The U. S. Department of Agriculture is still admonishing U.S. citizens to be wary of buying cans of food that are swollen, leaking, or dented, because they might contain the bacterium *Clostridium botulinum*.

However, the commercially canned product is not the real problem. Since 1925 there have been only four and possibly five deaths attributed to botulism poisoning from commercially canned foods, but some 700 deaths from botulism poisoning can be traced to improper home canning (20).

Therefore, the emphasis in this chapter will be on the preservation of foods in the home.

The various ways of preserving food differ, yet they all prevent the spoilage of our foods by one or more of the following ways:

(1) Asepsis, or the keeping out of microorganisms.
(2) Removal of microorganisms.
(3) Maintaining an oxygen-free, sealed environment.
(4) Use of high temperatures.
(5) Use of low temperatures.
(6) Drying.
(7) Use of chemical preservatives.
(8) Irradiation.
(9) Mechanical destruction of microorganisms (by means of grinding or the use of pressure).
(10) Combination of two or more of these methods.

No matter which method of food preservation we use, one danger exists which we may not be aware of. Fruits and other high-acid foods may react with the chemicals in galvanized metal pans. That chemical reaction can poison us. For this reason, it is better to use aluminum, plastic, or ceramic bowls and containers when we are preserving foods.

The ingredients in foods which bring about deterioration are water and substances such as proteins, carbohydrates, fats, minerals, and enzymes.

When we preserve foods, we utilize one or more of the following principles:

(1) Prevention or delay of microbial decomposition.
    a. Keeping out microorganisms (asepsis).
    b. Removal of microorganisms (filtration).
    c. Hindering the growth and activity of microorganisms by low temperature, drying, or chemicals.
    d. Killing the microorganisms through heat or radiation.

(2) Prevention or delay of self-decomposition of food.
  a. Destruction or inactivation of food enzymes (blanching).
  b. Prevention or delay of purely chemical reactions (prevention of oxidation by means of an antioxidant).
(3) Prevention of damage by insects, animals, or mechanical causes.

ASEPSIS

Even though the word asepsis may be strange and unfamiliar, Mother Nature has been practicing it for eons. Anytime there is a protective covering around food, microbial decomposition is delayed or even prevented. So Mother Nature has effectively provided protection with the shells of nuts, the husks of ear corn, the skins of fruits and vegetables, and the skin, membranes, or fat on meat and fish.

It is only when the protective covering has been damaged or decomposition has spread from the outer surfaces that the inner tissues are likely to decompose.

A plasticized cardboard package of frozen green peas in our supermarket is an example of how industry has capitalized on what Mother Nature has practiced for longer than man has been on earth.

REMOVAL OF MICROORGANISMS

This method of preserving foods is not too effective, because we just cannot sit down and pick the microorganisms off a piece of food. Removal can only be accomplished by filtration, centrifugation, washing, or trimming.

Filtration is the only successful method of completely removing organisms, but its use is limited to clear liquids. Centrifugation, or sedimentation, is not really effective because not all the microorganisms can be removed.

Washing is helpful, but, unfortunately, it can also be harmful if the water is contaminated or if the increased moisture encourages the growth of more microorganisms. Trimming off spoiled portions of a food or throwing away spoiled pieces or parts is helpful, but only for short periods.

Removal of microorganisms simply cannot guarantee that long-term storage of food will be successful.

PRESERVATION BY USE OF HIGH TEMPERATURES

Bacteria and enzymes are the most important causes of food spoilage. Yet we know that yeasts, molds, and oxidation also play a part.

Microorganisms are present in the water we drink, in the air we breathe, and on all objects under normal circumstances. As a result, foods become contaminated very easily.

We know that a single bacterium can produce millions of other bacteria in a matter of hours; in fact, we understand now (with a shiver up our spines) that bacteria can reproduce themselves more rapidly than any other form of life.

Since we are so constantly threatened by the possibility of overwhelming numbers of microorganisms in our foods, it's comforting to know we can destroy most of them with heat. (We can also destroy molds and yeasts, and inactivate enzymes with heat.) Heat destroys microorganisms by coagulating their protein.

There are several different ways we can use heat to destroy microorganisms and inactivate enzymes. The one we choose depends upon the kind of organism we are trying to destroy, its state of being, the kind of food being preserved, and the environment we place the food in during heating.

Pasteurization

Pasteur's method of preserving foods kills part, but not all, of the microorganisms. It usually involves using temperatures between 161° and 212° F. by steam, hot water, dry heat, or electric currents. The foods are then cooled immediately after the heat treatment.

Pasteurization is used under these circumstances:

(1) When more rigorous heat treatments would ruin the quality of the food.
(2) When the primary aim is to kill disease organisms.

(3) When the spoiler we're trying to destroy is not very heat-resistant.
(4) When any surviving microorganisms will be prevented from multiplying by additional preservative methods, such as the refrigeration of pasteurized milk.
(5) When the organisms which are competing for supremacy are to be destroyed, so that fermentation can take place, usually by adding a starter organism, as in cheese or yogurt making.

Canning

It's difficult for us to fully appreciate the impact of what Appert did for us when he developed the canning process. However, if we had to subsist for only one winter on the Dark Ages regimen of salted meat and root crops, we would probably make him an international hero.

The spoiler we must worry most about when we can food is *Clostridium botulinum.*

Spores of these bacteria are harmless by themselves, and they are very common in soil. Boiling will make the toxin harmless but cannot kill the spores which produce the toxin. These spores are extremely resistant to heat. We can kill them in a reasonable length of time by cooking them under pressure at a temperature of 240° F. (At high altitudes, the pressure must be increased in order to obtain this temperature.)

All of us have eaten botulism spores on fresh fruits and vegetables, but they pass harmlessly through our digestive tracts. The trouble occurs when contaminated food is stored in a sealed can or a jar, because the bacteria can grow only in the absence of air and usually only in low-acid foods. A pH of 4.5 or lower will usually prevent toxin production in most foods.

These low-acid foods are all vegetables (except tomatoes and rhubarb), seafood, meats, poultry, and fresh water fish. Bacteria do not grow well in more acid-like foods, such as tomatoes and rhubarb, fruits, or in foods like pickles which have been acidified to a pH of 4.6 or lower.

Another reason vegetables are prime targets for botulism is that they are grown close to the soil and, of course, *Clostridium botulinum* comes from the soil.

We should never attempt to can any food until we thoroughly understand the principles involved in the various methods used. A simple mistake can result in a fatal case of botulism.

There are three types of U. S. Department of Agriculture-approved methods for canning foods at home:

(1) Pressure canning.
(2) Boiling-water bath.
(3) Open-kettle method.

We should always consider heat, time, and food type when we decide which canning method to use. Two factors determine how much heat is needed, and how long it should be applied:

(1) The kind of substance or organisms that causes spoilage in the specific food.
(2) The natural or added acid in the food itself which aids in controlling the growth of bacteria.

Pressure canning is absolutely necessary for processing all foods that are low in acid. This means that all meats, seafood, and poultry, and all vegetables, except tomatoes and rhubarb, must be processed in a pressure canner. It takes more than just boiling temperatures to destroy botulism-causing organisms when they are present in low-acid foods.

The foods are packed, either raw or partially cooked, into jars, then sealed and cooked under pressure.

At 10 pounds of pressure, the temperature reaches 240° F. This temperature will destroy botulism organisms.

Boiling-water bath canning will destroy many other bacteria as well as yeasts and molds. This method is preferable for processing fruits, tomatoes, rhubarb, pickles, and relishes. The jars are sealed and covered with boiling water, then processed for specific lengths of time according to the types of food being processed, the size of the jars, and the altitude. Food can be packed into jars either raw or partially cooked.

In open-kettle processing, foods are merely boiled until done, then poured into sterilized jars and sealed. Today, the open-kettle method is recommended only for jams, jellies, and preserves.

In each of these methods, only tested, specific recipes should be followed. We should never, never rely on guesswork when we can foods.

About thirty or forty years ago many cookbooks still gave recipes for canning low-acid vegetables with the open-kettle method, and during World War II many home canners even used an oven-canning method for fruits and vegetables.

But the U. S. Department of Agriculture warns against these practices (11). Low-acid foods favor the growth of *Clostridium botulinum,* and the boiling temperatures which are used in the open-kettle method or the boiling-water bath method cannot kill the heat-resistant bacteria. Even though some home canners may get by with processing foods like corn and beans in the old wash boiler, it's as dangerous as a game of Russian Roulette. In oven canning, the dry heat does not easily penetrate the foods, and the oven regulator cannot keep the temperatures constant (28).

In the United States we are extremely fortunate to have a well-organized, widespread army of service-oriented specialists in the state university extension services.

Long active in the rural areas of the United States, the state extension cooperative services are arms of the U. S. Department of Agriculture and have remained forceful in farm activities.

But they have also followed us to the major urban areas, where they provide us with up-to-date information and expertise in the area of food preservation.

As our problems multiply and the food supply dwindles in the decades to come, these experts will be called upon more and more to contribute their know-how in many areas of food preparation and preservation.

Serious, even life-threatening, mistakes can be made in canning. Our greatest concern should be with the pressure canning of low-acid foods, since they are the prime targets of botulism.

If the dial gauge on a pressure canner is not working properly, we are defeated even before we begin.

Almost any state extension home economist in almost any county in the United States can provide us with the name and address of a specialist who can check the gauge on a pressure cooker. We should have it checked once each canning season, or even more often if we can foods a great deal. If the gauge is off 5 pounds or more, we should discard it. If the difference is less than 5 pounds, we can adjust. As a reminder, we should tie a tag on our canners stating the proper readings to use to get the correct pressure.

Since the food has to be processed at 10 pounds pressure (in order to achieve a temperature of 240° F.), this adjustment table should be used if the gauge varies up to 5 pounds in either direction.

If gauge reads too high:

    1 pound high—process at 11 pounds
    2 pounds high—process at 12 pounds
    3 pounds high—process at 13 pounds
    4 pounds high—process at 14 pounds

If gauge reads too low:

    1 pound low—process at 9 pounds
    2 pounds low—process at 8 pounds
    3 pounds low—process at 7 pounds
    4 pounds low—process at 6 pounds   (25).

Another precaution we should take at the beginning of each canning season is to clean the valves on our pressure cookers. We should draw a string or a narrow strip of cloth through the petcock and safety valve to make sure they are not clogged.

There are other potential trouble spots in the canning process, too. We doom ourselves to spoilage problems if we start with fruit or vegetables which are not perfect and clean. Produce is no better after canning than when it was picked and washed. Imperfect fruits and vegetables may spoil the rest of the food in their containers after processing.

When we wash fresh produce to prepare it for canning, we should cleanse it under running water or put it through several rinses. If we use the rinse method, we should lift the produce from the water as we rinse it, so the soil-related bacteria will drain off with the water. Only small batches should be cleaned, blanched, and processed at one time.

Food can be packed into jars or cans either raw or precooked. Most vegetables can be packed raw if they are processed in a pressure cooker. Most fruits can be packed either hot or cold; however, apples and rhubarb are best if we blanch them first.

Only jars which can withstand that all-important temperature of 240° F. should be used. This rules out our peanut butter and mayonnaise jars. Furthermore, we should use only jars which are specifically made for canning, and they should be perfect. No nicks or cracks. The rings should be perfect, and, of course, the lids should be new.

When we use a pressure canner at high altitudes, we should add ½ pound to the pressure gauge for each additional 1,000 feet in altitude. For instance, if processing requires 5 pounds pressure at sea level, we should use seven pounds at 4,000 feet, 9 pounds at 7,500 feet. If 15 pounds is needed at sea level, we should use 17 pounds at 4,000 feet, and 19 pounds at 7,500 feet.

When we use a boiling-water bath at high altitudes, we must make other changes. If the total processing time is 20 minutes or less, we must add a 1-minute increase for every 1,000 feet above sea level. If the total processing time is more than 20 minutes, we must add 2 minutes per every 1,000 feet above sea level (11).

But no matter where we are canning, the cardinal rule is to make sure the processing time is correct. No matter how carefully we have selected and prepared the produce, if it is not processed long enough, it will spoil.

If we can meat, poultry, or seafood, strict control of microorganisms must be even greater. Meats, fish, and poultry are all likely to be contaminated to begin with. Since they are low-acid, they provide an excellent medium for those bacteria to grow in, to say nothing of providing ideal conditions for *Clostridium botulinum* to grow and thrive. Therefore, all meats, poultry, seafood, or fresh water fish must be processed under pressure.

If refrigeration is not available, and if the maximum daily temperature is below 40° F., the meat may be processed as soon as the body heat is gone, usually within 6 to 24 hours (25).

Meats may be precooked or packed raw, but the meat must be from healthy animals and fowl or from fish or seafood taken from uncontaminated waters. Only fresh fish should be canned, and these should be bled thoroughly, then cleansed of all waste material. In addition, fish must be soaked in brine (salt water) for an hour before processing.

There is always a strong risk of botulism if short cuts are taken in canning meats, poultry, or seafood.

If meat must be held longer than a day or two before canning, it should be frozen at temperatures lower than 0° F. When we are ready to can, we may cut or saw the frozen meat into pieces of desired size. If frozen meat is thawed before it is canned, it should be thawed in a refrigerator at temperatures below 40° F. until most of the ice crystals have disappeared (25).

All meat, poultry, and fish must be kept clean and sanitary. Poultry must be rinsed well in cold, running water. All meats must be kept as cool as possible during handling, and all steps should be done quickly.

To keep down bacteria, utensils and work surfaces must be kept clean. We should scrub pans in hot soapy water and then rinse well with boiling water before putting meat into them. Knives and kitchen tools must also be washed in soapy water, then rinsed well in boiling water.

During the canning procedure, we should scrape surfaces of cutlery boards, wooden utensils, and work surfaces when necessary, then scrub with hot, soapy water and rinse with boiling water. We should then disinfect clean surfaces by using a liquid chlorine disinfectant diluted according to directions on the container. Wood surfaces should be cleansed with this solution, then left for 15 minutes and washed off with boiling water.

If we are unfortunate enough to have our home-canned food spoil, we can fix the blame (if we're still around) on one of the following reasons:

(1) The food was not sterilized properly, either by insufficient heat during canning or the pressure was not maintained at 10 pounds.

(2) We failed to wipe the sealing edge of the jar clean before placing the lid on it.
(3) The air was not sufficiently exhausted from the pressure canner before the regulator was placed over the vent.
(4) Water in the water-bath canner was not kept at a full boil throughout the entire processing time, or the water level dropped down below the top of the jars.
(5) The jars were nicked or cracked or had sharp sealing edges. (Jars manufactured for the specific purpose of canning have flat sealing edges.)
(6) In the open-kettle method, we filled a number of jars at one time and allowed the jellies, jams, or preserves to cool and become contaminated before sealing them with paraffin and/or caps and rings.
(7) The methods recommended by the manufacturer for tightening or sealing individual jars, caps, or lids were not followed.

Our concern and precautions must not end with the processing of home-canned foods.

Metal tops of jars should be tested within a few hours after canning by tapping them lightly with a metal spoon or knife. A clear, ringing sound means the seal is safe.

A dull, hollow sound means the seal is imperfect. If it has been longer than 5 hours since canning, we should throw the food away. (If less than 5 hours we can chill the food and use it soon.)

Jars should be labeled, dated, and stored in a dark, cool place where the temperature is between 45° and 60° F.

Still further precautions must be taken before we ever taste any low-acid home-canned food. Before we open any jars or cans of home-canned foods, we should always examine them to see if there has been any swelling, bulging, or seeping.

Any can that shows any damage of this sort should be discarded at once and not even opened.

If the can we want to open looks sound, we should still carefully examine the contents to be certain there is no foaming, no spurting of liquids, no molds, murky appearance, off color, or off odor.

We should never under any circumstances eat food with these characteristics.

The most important rule to remember when we are dealing with home-canned foods is that we must never, never taste low-acid foods until they have been boiled in an open kettle for 20 minutes. Boiling temperatures sustained for this length of time will destroy botulism toxin, even though it will not destroy the spores. But the spores do not harm us.

If the food develops an odor during cooking, it should be discarded. Cooking brings out spoiled odors, especially in canned meat.

Any food we suspect of being spoiled should be destroyed immediately so that not even an animal can eat it.

PRESERVATION BY USE OF LOW TEMPERATURE

If we think refrigeration of foods is the brain child of technology, we need to remember the Roman emperors kept their fresh foods in excellent condition with ice brought down by runners from the mountains.

Cooling through evaporation was accomplished in earthenware jars as early as 2500 B.C. in Egypt and India.

Although freezing as a means of preserving food is relatively new in the developed countries, it has been a common practice in arctic regions for thousands of years.

Most life processes are slowed down or stopped by low temperatures. This is not difficult for us to understand when we remember how stiff and cold we were when we stood in the ski-lift line for an hour at subzero temperatures, or how much we hate to crawl out of bed for a glass of water on a winter night.

The activity of microorganisms and enzymes is also slowed down by low temperatures. Bacteria die rapidly between 30° and 23° F., usually from starvation, but there are some kinds of bacteria which can survive in frozen foods for several years. Sugar, salt, protein, colloids, and fats protect microorganisms from extremely low temperatures, but high moisture and a low pH will hasten their death.

PUTTING UP FOOD

Freezing usually reduces the number of living organisms in foods. We certainly cannot expect freezing to have the sterilizing effect that heat achieves. This is a misconception many of us have; we think if we freeze meat or vegetables they are automatically safer and freer of microorganisms than they were before we froze them. Far from it. Thawing sets up just the right conditions for microorganisms to begin growing and multiplying at a rapid pace.

Cellar Storage

The old-fashioned root cellar didn't have an avocado green door on it, and there weren't any sliding shelves. However, that old-fashioned cellar utilized four basic and fundamental rules for keeping vegetables:

(1) Darkness.
(2) Temperature around 32° F.
(3) High humidity (about 85 percent).
(4) Good ventilation.

Cellar storage used to be far more successful an operation than it is now. Before we had centrally heated homes and warm basements, it was not difficult to keep a damp, dirt-floored section of the cellar or basement at the desirable storage temperature (somewhere between 32° and 40° F.). However, under correct circumstances, all root crops, celery, cabbage, apples, and some other similar produce items can be kept very successfully for limited periods in cellar storage.

Let's say we are faced with a situation where we must keep vegetables for as long as we can, without benefit of basement or refrigerator. A root cellar would be our solution.

The rules for root-cellaring have been lost to most of us, yet we are the grandchildren and great-grandchildren of pioneering Americans who understood root cellar storage very well.

This kind of storage is awkward and it should be attempted only in the areas of our country where the winter temperature does not drop below 28° F. or does not get above 55° F.

A shallow pit should be dug and lined with straw or leaves. Vegetables should be stacked in piles (keep each kind together). There should be straw between each layer of vegetables. The en-

KEEPING FOOD SAFE

tire cache should be covered with straw, and finally with about 6 inches of soil. (The straw must protect the vegetables from direct contact with the dirt.) Then the soil must be pressed down and covered with a tarp to make it waterproof.

Chilling

Chilling storage is accomplished at temperatures not far above freezing. There was a time when the iceman drove a team of horses and a wagon down the street and hauled ice from door to door. Although less colorful, our modern refrigerators and freezers are far more efficient, IF we use them correctly.

Enzymatic and microbial changes in foods are not prevented at chilling temperatures, but they are slowed down considerably. The ordinary refrigerator holds foods at temperatures between 32° and 40° F.

The old iceboxes the iceman used to service kept foods between 40° and 50° F. At one time, the recommended chilling storage temperature was somewhere below 50° F., but recent research has shown pathogens and spoilage agents can grow at these temperatures.

Since some pathogens (specifically *Staphylococcus aureus*) have been found growing at 46° F., modern refrigerators have been made to keep food considerably cooler.

Temperatures are not all we must be concerned with when we store foods at chilling temperatures. Fruits and vegetables lose weight and wilt when they are stored under conditions that are too dry. And we have learned that if humidity is too high, microorganisms are encouraged to grow.

The relative humidity and the temperature for chilling storage vary with the food stored. For example:

Product	Storage Temperature	Recommended Relative Humidity, Percent
Apricots	31–32°	85–90
Bananas	53–60°	85–90
Cabbage, lettuce	32°	90–95

Product	Storage Temperature	Recommended Relative Humidity, Percent
Nuts	32–36°	65–70
Onions	32°	70–75
Tomatoes (ripe)	40–45°	85–90

It had long been a most perplexing problem, this business of holding foods at just the proper temperature and humidity, but in 1834 an American named J. Perkins invented the first refrigeration machine. The first practical compressors were invented between 1874 and 1876 (8).

Smugness must have been a household vice in the late 1800s and early 1900s. Suddenly, for the first time in history, man could shut winter up inside a little box and store his foods as the Eskimos could in the snow.

The refrigerator brought about revolutionary changes in the way man could store his food. As a matter of fact, the refrigerator caused a revolution in eating habits. Suddenly, we could indulge in ice creams and molded salads at our leisure.

Freezing

We can scarcely visualize modern life without some kind of freezer in our homes. Whether it is no more than an ice cube compartment in our refrigerator or a sleek, yawning deep freeze, freezers have undoubtedly played their part in helping woman, whether she is a household executive or a career-minded individual, or both.

Whole meals and main dishes can be cooked or bought and frozen for later use.

Sometimes we take our freezers so much for granted, we begin to think they can perform miracles. The fact remains that frozen food can be no better than the food was before it was frozen. We must take care to freeze only good quality, tender fruits and vegetables, and only wholesome meats, fish, and poultry.

We cannot remind ourselves too often that the freezing process does not kill all food-spoilage organisms; it merely stops their

KEEPING FOOD SAFE

growth and multiplication. As soon as food is thawed, microorganisms begin to grow and multiply again. For this reason, we must make sure the numbers of bacteria on food we freeze are kept at a minimum.

It is essential that we follow strict sanitary measures when we freeze foods. The food and everything it touches must be clean. Foods to be frozen should be handled as little as possible.

We should use the same care when we prepare vegetables for freezing as when we prepare them for canning. Fresh produce should be washed under cool, running water, or it should be washed through several rinses. If we use the rinse method, we should lift produce out of the rinse water to allow microorganisms to drain off with the water.

Since the enzymatic activities are not stopped but merely slowed by freezing, most vegetables should be blanched (precooked) before they are frozen. This stops most enzymatic activity. Enzymes cause ripening, which will in time cause spoilage. Blanching also helps reduce the number of microorganisms on the food and wilts leafy vegetables so they will pack down better.

We should never use copper or iron utensils for blanching. A large aluminum or unchipped enamelware kettle is better. We should always follow a recipe when we are blanching a particular vegetable, because different types of foods require different lengths of blanching time.

The water in the blanching kettle should be boiling rapidly before the vegetables are added, and it should keep right on boiling through the entire blanching period.

It is best to put about a pound of prepared, cleaned vegetables in a collander and then lower it into about 4 quarts or more of rapidly boiling water. The kettle and collander should then be covered tightly, and the heat should be left on high. Only then should we begin timing the blanching process.

When the blanching period is over, the container of vegetables should be removed from the boiling water and plunged immediately into a large pan or sink of very cold or iced water. The vegetables should be left in the cold water twice as long as they were left in the blanching kettle.

PUTTING UP FOOD

The vegetables which freeze best are those which are grown above the ground and which are cooked before being eaten.

Fruits, like vegetables, should be perfect, or nearly so, when we freeze them. Ascorbic acid (vitamin C) will have to be added to some fruits to keep them from turning dark and to help preserve natural fresh flavors.

As with vegetables, we should always refer to specific recipes for freezing fruits. Fruits can be frozen by either the dry pack or syrup pack methods. Generally, however, we should peel, pit, or otherwise prepare the fruit, following recipe directions. If fruit is to be packed in syrup, we should use the recommended type and strength of syrup. The acid solution should be prepared immediately before use.

The fruit should be packed in rigid plastic containers or plastic bags, then if syrup is called for, it should be added at this point. Proper headspace should be allowed, and container lids must seal securely. Twist closures must be securely fastened on plastic bags.

Fruits should be thawed unopened, because they darken and get soft quickly. Thawing 1 pint of frozen fruit takes about 30 minutes to an hour in cold water, or 2 to 4 hours at room temperature, or 5 to 8 hours in the refrigerator. Refrigerator thawing produces the best results.

When we freeze meat, we should rewrap the store wrap with freezer paper, heavy-duty foil, or heavy plastic wrap. Meat cannot be frozen safely for lengthy periods unless it is kept at 0° F. or less. It should be divided into individual, family, and company-sized portions.

Large, frozen cuts of meat should be completely thawed before we cook them; otherwise, the outside will be cooked and the inside will be barely warm, hardly a safe situation. We should always use a meat thermometer when we are cooking large portions of meat so we'll know what's going on inside.

We should never freeze home-stuffed poultry.

Seafood and fresh water fish pose different problems. Lean fish should be dipped for 20 to 30 seconds in a cold-water brine solution made by dissolving ½ cup salt in 1 quart water. High-fat content fish such as bonito, butterfish, halibut, herring, mackerel, lake trout, pompano, and salmon should not be brine-dipped.

Several small fish, steaks, or filets may be placed in a loaf pan, wax carton, or coffee can, covered with water and frozen. When the blocks are solidly frozen, they can be removed and wrapped in freezer wrapping materials for storage. Whole fish can be repeatedly dipped in near-freezing water until a glaze has formed; then the fish should be wrapped.

To freeze whole eggs, the whites and yolks should be gently mixed with a fork, then 1 teaspoon salt or sugar added for each 2 cups of eggs. To freeze yolks separately, they should be mixed thoroughly with 2 teaspoons salt or sugar for each 2 cups yolks. Nothing needs to be added to egg whites for freezing. Eggs should be placed in rigid containers and ample headroom left before sealing. Containers should be dated and the addition of either salt or sugar noted. They should be thawed in cold water and used within 12 hours after thawing. Two tablespoons of thawed white equal 1 egg white; 1⅓ tablespoons thawed yolk equal 1 egg yolk, and 3 tablespoons thawed whole egg equal 1 whole egg.

Home-baked bread should be baked, allowed to cool, then wrapped in freezer wrap. Bakery bread should be over-wrapped. All quick breads freeze well, and they should be wrapped in plastic or foil. We should thaw breads in their wrappers at room temperature or heat in a 325° F. oven 15 to 30 minutes, depending on size.

Combination dishes to be frozen should not be overcooked. Those which we have just cooked and want to freeze should be cooled as quickly as possible by placing the container in ice water. Then the dish should be sealed or wrapped tightly and frozen immediately.

Since the cold, circulating air in a freezer tends to dry out food, we must be careful to seal all packages very carefully. Any containers or materials which are used to package foods for freezing should be:

(1) Airtight, because the chemical action of oxygen creates peculiar flavor changes in frozen foods. Exposure to

air also causes a dried-out, seared texture and appearance known as "freezer burn."
(2) Vaporproof, because tastes and odors can be transferred from food to food if wrappings can't keep them in. That's why proper freezer containers and wraps are all nonporous (sealed-surface) materials. Nothing can get in or out.
(3) Moisture-proof, because the air held inside any freezer is damp, as well as cold. Dampness can cause poorly packaged food to be inferior since the food is exposed to both freezer burn and undesirable flavor/odor combinations.

Even if we successfully freeze food, we can still ruin it by incorrect thawing. Just exactly what happens when food thaws? As the ice crystals melt, the liquid either is absorbed back into the tissues or leaks out from the food. If we thaw food slowly at cool temperatures, we'll be far more likely to have a better, more moist food. Surviving enzymes will begin to be active again during thawing but, if we cook the food immediately, action is stopped.

We run into trouble with thawing when we allow frozen foods to thaw at room temperatures. Dangerous spoilage can occur in 3 or 4 hours.

Few of us realize that if precooked frozen foods are kept at room temperatures too long after thawing, there can be growth and toxin production by either staphylococci or *Clostridium botulinum*. Unfortunately, the final warming up of precooked dishes in many home kitchens and restaurants is not always sufficient to appreciably reduce the numbers of organisms present or to destroy their toxin.

There are important but little-known rules for heating frozen dishes. These rules will assure that no microorganisms will be given a chance to grow or produce toxin:

(1) When we heat combination dishes, we should take the container directly from the freezer and put it into a hot oven. (This is why freezer-to-oven casseroles are so valuable.)

KEEPING FOOD SAFE

(2) If it is necessary to partially thaw food before removing it from non-ovenproof rigid containers (in order to remove cellophane between layers or to remove food from the container), we should place the package in lukewarm water for a few minutes first. If necessary, we can allow the food to stand in the container at room temperature for 30 to 60 minutes. It may also be necessary to thaw food slightly if we are going to cook it over direct heat.

(3) We should reheat frozen soups and other dishes that contain a high percentage of liquid in a saucepan over direct heat. The heat should be kept low, and the food should be stirred gently, just to prevent sticking.

(4) We can reheat some dishes in a double boiler as well as in a hot oven. (However, reheating in the oven takes little or no attention, though it takes more time.)

(5) If we reheat food in a double boiler, we should start with warm water in the bottom pan. The frozen mass should be broken up as it thaws and stirred gently only when necessary to keep the food from sticking. (Too much stirring may ruin the texture of the food.)

(6) Larger blocks of frozen food or several smaller blocks will heat faster in a wide-bottom saucepan set in another pan of hot water.

(7) If we prefer to thaw food completely before reheating, it should *always* be thawed in the refrigerator.

(8) In general, a temperature of 300° F. will heat the following amounts of frozen food in the specified amounts of time:
   a. Pint of frozen food: allow 1 hour.
   b. Quart of frozen food: allow 1 hour, 45 minutes.
   c. A few minutes should always be added at high altitudes.

(9) We should always make sure any reheated frozen food is hot all the way through and is bubbling.

(10) The directions on packages of commercially frozen foods must be followed exactly in order to insure that

food will be safe. (Extra time should be allowed in high altitudes.)
(11) We can heat frozen meat, poultry, or fish without thawing, but we must allow more cooking time to be sure the center is properly cooked. We should allow at least 1½ times as long to cook them as would be required for unfrozen products of the same size and weight.

What if we have thawed a combination dish properly, but Sally fell off her bike, broke her arm, and had to be taken to the emergency room at the hospital? To save time and effort we bought hamburgers for the family on the way home. Should we refreeze that leftover casserole?

Absolutely not. Combination dishes such as meat, poultry and fish pies, stews, hash, etc., should never be refrozen under any circumstances.

Occasionally, frozen foods are partially or completely thawed before we discover the power has been off and the freezer is not working. Is it safe to refreeze the foods we had stored in the freezer?

That depends on the temperature at which the foods were held, the length of time they were held after thawing, and the type of food.

We may safely refreeze any foods except fish and combination dishes if they still contain ice crystals and have been held no longer than 1 or 2 days at refrigerator temperatures after thawing.

However, no *completely defrosted* foods, except strong acid fruits, should ever be refrozen. Defrosted low-acid foods, if refrozen, are possible sources of food illness. We need to remember that defrosted low-acid foods may be spoiled, even if they have no telltale odor.

Acid fruit juices will not support microbial growth, but they will ferment. However, if the flavor is flat, even juices should be discarded.

We should never refreeze partially thawed foods which we have kept in the refrigerator for more than 48 hours. Common sense tells us that any off-color or off-odor thawed meat, fish, seafood, or poultry should not only never be refrozen, but should be discarded entirely.

No thawed vegetables should ever be refrozen—we must remember they contain soil bacteria.

What about the ice cream we've just bought that thaws when we made an extra stop at the garden shop on an April afternoon? If we spot a colorful Mexican pottery flowerpot that would look just right on the patio, that is all well and good, except that the ice cream is slowly melting in the trunk of the car. What should we do with that softened, runny ice cream when we get home? Unfortunately, it must be thrown out. We should never refreeze ice cream.

If we suddenly find our freezer is not working and foods are defrosted, these are the questions we should ask ourselves:

(1) Which foods can be stored in the refrigerator and eaten within a day or two?
(2) Which foods must be discarded?

All vegetables must be discarded, because they may contain botulism spores. They are packed in airtight packages which provide ideal conditions for toxin production. Very cold (below 40° F.) foods which do not contain ice crystals can be refrigerated or used immediately.

We must be extremely conscious of the temperatures at which we hold foods, because between 44° and 125° F. there is increasingly rapid growth of bacteria. However, below 44° F. and down to 10° F., bacterial growth is slowed considerably. Below 10° F. there is no microbial growth. (See Table 6.) If we lower the temperature in our freezers to −10° F., the storage time of most frozen foods can be doubled. (See Chapter 8 for specifics about storing frozen foods.)

Freezer temperatures should be maintained at a constant level, because fluctuating levels of coldness tend to dry out foods and make them more susceptible to the development of off odors.

We should never attempt to freeze food in the ice-cube compartments of one-door, conventional refrigerators. These appliances were not designed to bring food temperatures down and

PUTTING UP FOOD

maintain them at that all-important and critical 0° F. point. (See Chapter 8 for detailed instructions about holding frozen foods in the ice-cube compartments of conventional, one-door refrigerators.)

PRESERVATION BY USE OF DRYING

Mother Nature herself taught primitive man to dry foods. It must not have taken our cave-dwelling friend long to learn that the seeds pulled from wild wheat kept all through the long winter. All the time the seeds were maturing, nature was removing one of the main causes of deterioration, water.

As long as seeds are kept from moisture, they will keep well. The explorers who first searched through Egyptian ruins found well-preserved kernels of wheat.

Ancient civilizations in arid regions of the Near East undoubtedly depended upon dried fruits as staples. At first man probably found the dried fruits under the trees, but later he must have decided he could help matters along by picking fruits and nuts at their prime and placing them out in the sun to dry. Eventually, he developed a system of drying fruit in hot sand.

Meat was dried in arid regions by beating it with rocks and then leaving it exposed to the sun and hot desert winds. The Scandinavians later discovered that cold air and a brisk north wind did almost as good a job of drying as the desert atmosphere did.

Drying is not only the oldest way known to preserve food, it is also the simplest. Moisture can be removed from foods by any number of methods, from the ancient practice of drying in the sun to modern commercial ways.

The Swiss lake dwellers learned to dry fruits and berries and store them in earthenware vessels. Early Egyptians salted and sun-dried fish and fruits, and they even constructed buildings to protect them. The early Chinese preserved fish and meat with salt, and the early Romans dried asparagus and cabbage. Dried figs even served as bread substitutes in some areas in and around Rome (8).

Freezing in conjunction with drying is one of the most ancient food preservation methods used by man. The pre-Columbian Indians, who lived in what is now Peru and Bolivia, were able to dry and store their native potatoes by freezing them at night, then thawing them and trampling the potatoes the next day, squeezing the juice out with their feet. The sun dried the material which was left, and the potatoes kept nearly as well as our modern-day instant potatoes (8).

When refrigeration and canning were developed as feasible means of preserving foods, drying methods were forgotten by many people, and much of the folklore surrounding the methods was forgotten. For this reason, drying has never received the attention and research that the other means of preserving have.

It has only been in recent years, since we have begun to anticipate a prolonged fuel shortage coupled with impending food scarcities, that an interest in drying has been revived.

Drying preserves food in two major ways:

(1) Water, which is necessary for the growth of microorganisms and enzymatic activity, is removed.
(2) With the removal of water, the concentration of sugars and acids is increased, creating a chemical environment which is unfavorable to the growth of many microorganisms.

Dehydration does slow down enzymatic activity in foods, but it does not prevent it entirely. Dehydration must usually also utilize one of two methods of enzyme destruction, either blanching vegetables or sulfuring fruits.

If we apply heat prior to the drying process, we can eliminate a lot of the microorganisms. Generally speaking, when heat is applied before or during drying, all yeasts and most bacteria are destroyed; however, spores of bacteria and molds usually survive, as do the vegetative cells of a few species of heat-resistant bacteria.

But if we are careful and follow directions during the drying process and storage conditions are adequate, the chances are quite good that we will have no significant number of organisms left on our dried foods.

The numbers of microorganisms on a piece of dried fruit can vary from a few hundred per gram to many thousands. Microorganisms on dried vegetables can range from very few to millions per gram. By now we know there is a greater potential for high numbers of bacteria on vegetables, because they grow in or near the soil.

Bacteria usually cannot grow if there is less than 18 percent moisture. Yeasts have to have at least 20 percent moisture, but molds need only 13 to 16 percent.

Drying can either be done outdoors in the sun or inside in an oven or dehydrator. Either method takes a good deal of time and requires our constant attention, especially at the beginning and end of the processes. The faster the process is, the lower the contamination and the better the quality of the food. Once the drying process is started, we should complete it as quickly as possible.

Under no circumstances should we ever attempt to dry food outdoors if we live in a smog belt, or in an urban area surrounded by superhighways. Even if we live less than 1,000 yards from a well-traveled secondary road, we should not consider sun drying. The pollution may dangerously contaminate the food (19).

Fruits and vegetables (dried either in the sun or in ovens or dehydrators) are usually dried on mesh-covered trays. In order to insure cleanliness, we should scrub the mesh or screen with an alkaline cleansing agent before using it. It's best to use a stiff brush and then rinse in clear water thoroughly.

We should never use metal screening, because copper will destroy vitamin C, and most galvanized screen has been treated with zinc and cadmium, some of the villains responsible for chemical poisoning (1).

Regardless of how thrilled primitive man may have been with the first dried fruits he discovered, we may be disappointed with our sun-dried foods. Foods dried in controlled heat will always be superior to sun-dried food in color, flavor, cooking quality, and nutritive value. The food will also be more likely to be freer of insects and insect eggs if it is dried inside.

The initial temperature in controlled drying is critical. If beginning temperatures are too high, the water-filled cells may expand and burst, or the food may harden on the surface, making it harder for moisture to escape. The moist interior then sets up a perfect environment for microorganisms.

Since most ovens cannot be set as low as the starting temperatures are supposed to be, the oven should be preheated at the lowest temperature possible, then allowed to cool down before the food is put into it.

Although some drying instructions recommend disconnecting the top unit in electric ovens before the drying process begins, experts with public utility companies frown on the practice. Rather, they recommend using only one rack in the oven, the bottom one, and putting it as close to the bottom of the oven as possible. Then drying instructions should be followed to the letter. Food should be stirred often. Most foods will require constant watching and frequent stirring near the beginning of the process.

If we are lucky enough to have a newer model range which will hold temperatures as low as 140° F., we'll have much better luck with drying in the oven.

It is absolutely essential to dry foods until the food springs back when we squeeze some in our hands. There should be no feel of moisture. Any moisture will encourage microbial growth.

All fruits, except dates and figs, should be washed before drying, and excess moisture should be removed with paper toweling. Dates and figs should be wiped off vigorously with a damp, rough cloth. Vegetables should be thoroughly washed and, of course, all produce should be young, tender, and completely free of insects.

Many fruits, particularly apples, apricots, peaches, nectarines, and pears, tend to darken during storage. Sulfuring preserves color as well as flavor and decreases the loss of vitamins A and C. It also prevents insect infestation.

To sulfur fruits, we should place the fruit, cut side up, on hardware cloth or screening on frames or trays. We should stack the trays to allow airspace for the circulation of sulfur fumes. Then the trays should be covered with a tight wooden box or heavy cardboard carton. The bottom tray should be about 10 inches above the burning sulfur to prevent scorching.

We should use 1 teaspoon sulfur per pound of prepared fruit if sulfuring takes less than 3 hours. If the sulfuring process used in a tested recipe calls for more than 3 hours of sulfuring, we should use 3 teaspoons sulfur (1).

We can buy the sulfur at almost any garden store or drug store.

It should be rolled loosely in a small piece of paper which we can twist at the ends to close. Then the roll should be put in an old pan or tin can; the paper should be set on fire and the trays covered quickly with the tight box, leaving the bottom of the box propped up several inches to allow for some ventilation. The box should remain over the fruit this way until the sulfur has burned completely; then we should remove the trays of fruit out into the sun or into the oven or dehydrator. Sulfuring should always be done outside.

When we expose the fruit to the sulfur fumes, we are not endangering our health when we eat the fruit. The heat of drying and the subsequent cooking will get rid of almost all of the sulfur.

Vegetables should always be blanched before they are dried. This not only helps salvage the vitamins, but it also sets the color and hastens the drying by relaxing the tissues. It also helps prevent undesirable changes in flavor during drying. Steaming is better than boiling, but we can use either method.

Meats should be cut into strips before they are dried. We should make sure all the fat is removed and the meat must be as thin as possible. We can coat the meat with salt and even use pepper if we have to dry the meat in the sun. (Pepper wards off insects.) The meat should also be covered with cheesecloth if the drying is done outside.

But we must remember that meats dried in the sun may harbor insect eggs. If we are ever suspicious that any dried food may have been infested, we can place it on trays in an oven for 3 minutes at 150° F.

Most dried food should be stored tightly sealed and kept in a dark, cool room or stored in the refrigerator. Once we open a container of dried food, we should store it in the refrigerator.

Tin cans or boxes with fitted lids, glass jars or moisture-vapor-proof freezer cartons or bags all make good containers for storing dried foods. We can seal lids by dipping them in hot paraffin, and plastic bags can be heat sealed.

Foods that are seemingly "bone dry" can be spoiled by reabsorbing moisture in storage. We should check dried foods from time to time during storage to see if they are staying dry.

Dried foods will usually remain stable as long as they are protected from water, air, sunlight, heat, and contaminants. For that reason, the packaging around them is critically important.

PRESERVATION BY USE OF PRESERVATIVES

When man discovered how to use fire for cooking, he no doubt also learned that meats and fish could be preserved by smoking. When meat is penetrated by smoke, its moisture content is lowered and the color and flavor are enhanced. The meat is also covered with small amounts of phenols and resins which furnish some protection from both bacteria and oxidation (9).

Man has preserved foods chemically for a long time. Next to drying, chemical preservation is probably the oldest form of holding food. Its beginning is lost in antiquity.

Preservatives are chemical agents which retard, hinder, or mask undesirable changes in foods.

A preservative will inhibit microorganisms by interfering with their cell membranes, their enzyme activity, or their genetic mechanism. Preservatives can also be used as antioxidants to slow down the oxidation of unsaturated fats, as firming agents and as stabilizers to prevent chemical changes. They can also be counted on to keep out many microorganisms. But with any preservative, there are generally some undesirable side effects.

Actually, man is still searching for the ideal preservative. The Federal Food, Drug and Chemical Act, as amended by the Food Additives Amendment of 1958, defines a chemical preservative as "any chemical which, when added to food, tends to prevent or retard deterioration thereof; but does not include common salt, sugars, vinegars, spices, or oils extracted from spices, or substances added by wood smoke" (9).

Many preservatives used today were discovered through trial and error in the ancient world. Some preservative techniques used today are still based on these experiences rather than on scientific investigation. Modern preservatives have been in use for less than two hundred years. Scientists are still not sure why some of the old-time methods work (8).

PUTTING UP FOOD

Pickles, olives, sauerkraut, and even fruit preserves are partially or wholly preserved by chemicals. When we make up a batch of pickles, the preservatives are salt and vinegar. Alcohol is a preservative in wines and liquors, and salt is the preservative in sauerkraut. The high concentration of sugar in jams and jellies also acts as a preservative.

Chemicals preserve foods by serving as a poison to the microorganisms or by providing an environment in which microorganisms cannot grow, even though it may not kill all of them.

PRESERVATION BY RADIATION

As a new food preservation method of great potential, ionizing radiation has advanced in only two decades from a laboratory curiosity to commercially feasible methods. Radiation was first approved by the Food and Drug Administration in 1963 for use on bacon (28).

Most pathogens and other microorganisms are destroyed by bombarding with ionizing particles or rays. Several kinds of rays have been tried, but gamma rays have certain advantages. They can penetrate farther into foods than can alpha or beta rays.

The major problems to be overcome before radiation can be used on a large-scale basis for food preservation are the high costs and undesirable side reactions on the food.

Storage time of fresh haddock at 32° F. can be increased with a radiation dosage of 250,000 rads from 6 to 9 days to 25 days (28).

This dosage destroys the bulk of the non-spore-forming microorganisms. But if the fish are treated in this manner and we let them sit out at room temperature, instead of the recommended 32° F., there would be danger that any surviving *C. botulinum* spores would grow and produce toxin. So even foods which have been treated with radiation also require specific storage temperature protection.

Radiation, one of the newest means of preserving foods, is but another step in mankind's continual efforts to put up foods in times of plenty against the times of need.

Table 6

TEMPERATURE OF FOOD FOR CONTROL OF BACTERIA

°F

- **250–240**: Canning temperatures for low-acid foods in pressure canner.
- **240–212**: Canning temperatures for fruits, tomatoes, and pickles in waterbath canner.
- **212–165**: Cooking temperatures destroy most bacteria.
- **165–140**: Warming temperatures prevent growth but allow survival of some bacteria.
- **140–120**: Some bacterial growth may occur. Many bacteria survive.
- **120–60**: DANGER ZONE. Temperatures in this zone allow rapid growth of bacteria and production of toxins by some bacteria.
- **60–40**: Some growth of food poisoning bacteria may occur.
- **40–32**: Cold temperatures permit slow growth of some bacteria that cause spoilage.
- **32–0**: Freezing temperatures stop growth of bacteria, but may allow bacteria to survive.

8

Stocking the Larder

We are all involved in the food business. Shopping for food on a weekly or monthly basis makes us at least part-timers in this multi-million-dollar enterprise.

Since most of us eat three meals a day (that adds up to 1,095 meals a year), we should be vitally concerned with the safest, most profitable and efficient ways to buy and store our food.

If we buy food which has spoiled or is on the borderline between wholesomeness and deterioration, we are wasting our time and money. Even worse, we are endangering the health and lives of our families.

We are luckier than our mothers and grandmothers were. We have efficient, dependable refrigerators and freezers which will store food safely for months. However, these appliances cannot work magic. The food we store in them can be no better or safer than the food we bring home from the market. The safety and wholesomeness of the food we store at home begins for us in the supermarket.

No buyer for a large chain of supermarkets would consider buying a carload of bulging cans or thawing frozen foods. However, defects in foods can occur after they have been purchased and are in transit to the store or while on display. This means that even though we can usually depend on our supermarkets to sell us fresh and wholesome food, we still must be alert and use our common sense.

Fortunately, more and more stores are using open dating and unit pricing to make it easier for us to make wise decisions.

Unit-pricing lets us compare the cost of different brands and package sizes of the same product. Shelf tags list the price per pound, gallon, or ounce.

Open dating assures us the perishable items in our supermarkets are fresh when we buy them. The rest is up to us. We must make certain to use perishable items before the expiration date.

We should carefully examine each and every item we buy to detect possible spoilage. We should never buy a package that is torn or imperfectly sealed. We must reject bulging cans and soft frozen foods.

Labels are all-important. They should be read carefully, because they describe the ingredients, special storage instructions, and serving directions.

Many consumers never stop to think that the frozen food display cases can be loaded incorrectly. We must check the load lines or frost lines on these cases. Any foods stored above this line may not be safe.

We must make sure that perishable products such as milk or raw meat and eggs are refrigerated in the stores, and that items like barbecued chicken and desserts in the "deli" are held at temperatures which will not allow bacteria to multiply.

Perishable food should be bought in small quantities. Frozen products should be the last items we put into our grocery carts before we go through the check stand.

The vast majority of food dealers are reliable, honest businessmen. Nevertheless, if we suspect that a law has been violated, we can and should report our suspicions to the proper authorities. (See Chapter 9.)

All the care and planning the food industry lavishes on keeping food safe will be wasted if we are careless.

The supermarket should be our last stop. Frozen foods, fresh meats, milk, eggs, and cheese can become hazardous if we don't keep temperature changes to a minimum. Fresh produce will wilt and begin to spoil if we leave it exposed to heat and sunlight in the back of a car for very long. If we are ever tempted to make an extra stop on the way home from the supermarket, we should remind ourselves that food spoilage is serious business.

Where we do our major food shopping is important. We should choose a store close to home which offers reasonable prices, good variety, and high quality.

There was a time when store-hopping was practiced widely. However, precious fuels can be wasted trying to save pennies at a store miles away. Generally, it is better to find a store we like and trust and then stay with it.

We should consider if the store has a good reputation, if it is clean, if the packages and cans are rotated, and if there is dust on any items. Does the store sell partially spoiled or damaged foods, dented cans, or partially frozen foods? Does the meat department look sanitary?

PART 1

How to Buy

The larger the size, the lower the unit price, usually. A glance at the unit price tag on the shelf below the item will confirm or deny this rule of thumb, but we should never take it for granted.

Even if the larger size is more economical, it may still be a bad buy for us if:

(1) Food is left over and we have to throw it away.
(2) Using it up means monotonous meals.
(3) We cannot store it properly, safely, and conveniently.

A large-quantity purchase can mean savings, fewer trips to the store, and added convenience, however.

MEAT

Whether we buy two T-bone steaks or a side of beef depends on how much we can afford to spend at one time, the amount of freezer storage space we have, and how much meat our family consumes. (One cubic foot of freezer space will usually hold 35 to 40 pounds of cut, wrapped, and frozen meat.)

If we decide to buy beef in quantity for our freezers, there are certain facts we need to take into consideration.

Animals have their differences—in thickness of muscling, amount of fat and bone, age and quality of their lean. USDA quality grades are a good measure of beef's tenderness, juiciness, and flavor. USDA yield grades (1 through 5) are a good measure of the proportion of usable meat in a carcass. Beef identified by both quality and yield grade is readily available.

If we decide on a 600-pound carcass, we should know the actual beef we will be able to put in the freezer will amount to only about 400 pounds.

As a rule, a 1,000-pound steer will yield a 600-pound carcass. After a steer is butchered, the packer sells the carcass to a retailer who trims away pounds of fat, bone, and waste (38).

If we decide on a side of beef, we should make these calculations after the beef has been purchased. Then we can more easily determine if there have been significant savings.

(1) The total cost for the side or carcass as it hangs on the hook.
(2) The weights of the various cuts as determined on a home scale.
(3) The cost of the take-home meat.
(4) The cost of the same cuts bought other ways. (We must not forget that we can sometimes buy the same cuts on sale.)

When we buy smaller cuts of beef for our families, we need to take into consideration how much the meat weighs and the number of people to be fed.

An average serving of meat is three ounces. We can generally count on 3 to 4 servings per pound from meat with little fat or bone; 2 to 3 servings per pound from cuts with medium fat or bone (most roasts, ham, chops), and 1 to 2 servings per pound from cuts with a lot of bone and fat (steaks and chops) (38).

There are so many different cuts of meat in today's supermarkets, we could prepare a different meat dish every day of the year.

The muscles and bones of all meat animals are similar, however. The tender cuts in beef are the same in lamb and pork.

Usually a beef carcass is cut into 9 wholesale cuts. Each contains muscles which vary in tenderness according to how much they were used by the animal. Because they did more work, motion muscles such as those in the shoulder, hip, and front and hind legs are less tender than the support muscles along the animal's back (22).

These 9 wholesale cuts divide the carcass into tender and less tender cuts. If we remember these, or at least refer to this list now and then, our meat shopping will be much easier:

(1) Chuck: Cut from the shoulder. (Blade chuck cuts from the shoulder blade area are slightly more tender than arm chuck cuts which come from below the blade area near the front leg. Cook with moist heat.)

(2) Rib: Cut from the rib section. (Includes rib steaks and roasts. One of the very tender cuts. Cook with dry heat.)

(3) Short loin: Cut from the middle of the back. (Usually very tender. Contains club, T-bone, porterhouse, strip loin, and filet mignon [tenderloin] steaks. Cook with dry heat.)

(4) Sirloin: Cut from the next section down the back. (Contains sirloin steaks. Usually slightly less tender than the short loin. Cook with dry heat for USDA prime, choice, or good grades.)

(5) Round: Cut from the hind leg. (Contains round steak, rump roast, heel-of-round roast, and sirloin tip roasts and steaks. Less tender than the sirloin section. Use moist heat for most cuts, but the rump roast and sirloin tip roast and steak can be cooked with dry heat if it is USDA prime, choice, or good grades.)

(6) Foreshank: Cut from around the front leg above the knee. (Used for stew meat and ground beef. Definitely a less tender section. Cook with moist heat.)

(7) Brisket: Cut from the lower chest section of the animal.

(Can be used fresh, but it is often cured and sold as corned beef. A less tender section; cook with moist heat.)
(8) Plate: Cut from just beneath the ribs from what we'd call the back portion of the chest. (Used for short ribs and beef for grinding or stewing. A less tender cut. Cook with moist heat.)
(9) Flank: Cut from what we'd call the animal's belly. (Yields flank steak and beef for grinding or stewing. Also a less tender section. Cook with moist heat.)

For many years the names of meat cuts have varied from locale to locale. We could walk into the store or meat market and come out with a chateaubriand and a filet mignon. We would have had two identical cuts.

To cut down on the confusion, the Industry-Wide Cooperative Meat Identification Standards Committee has proposed standardized meat cut names which are gaining wide acceptance. The system involves the following information:

(1) The animal the meat comes from.
(2) The part of the animal the meat is cut from.
(3) The specific portion or retail cut.

The regional or fanciful name (such as Spencer steak) may also be included but will usually appear in a parenthesis at the bottom of the label.

Even if we remember the information about the various cuts, we still need to know and understand the different meat grades.

The U. S. Department of Agriculture has established beef grades which give us an indication of tenderness and flavor. Five of the 8 grades may be available in retail stores:

(1) Prime: The best grade given by the USDA. Much of this is sold to restaurants.
(2) Choice: The grade most often found in supermarkets. This means the meat will be juicy, tender, and have a good flavor.
(3) Good: Not as juicy and flavorful as prime or choice,

STOCKING THE LARDER

but fairly tender with less waste, because it has little fat.

(4) Standard: This meat has very little fat and a mild flavor. It lacks juiciness but is relatively tender, because it comes from young animals.

(5) Commercial: Cut from older animals. This grade is not tender and needs long, slow cooking in moist heat to tenderize it (38).

Good-to-top quality beef usually has a moderate fat covering over most of the exterior. The lean should be firm, velvety-looking, and fine-grained. This quality of beef is produced from young animals. The bones are porous and red. Bones are white and flinty in older animals.

The lean of veal is a light, grayish pink. The meat is fine-grained, firm, and velvety, and there is only a small amount of smooth, firm white fat. The bones are porous and reddish.

The color of desirable pork will range from a grayish pink to a delicate rose. The meat should be fine-grained, firm, and free of excess moisture. The lean should be well-marbled with fat and covered with firm white fat.

Lamb varies in color from light to dark pink and in yearlings from medium pink to light red. The fat should be chalky white (38).

FISH AND SEAFOOD

There is no rule of thumb as far as quantity goes for buying fish. However, this table can serve as a guide:

(1) Whole or drawn fish: yields ½ to ¾ pound per portion.
(2) Dressed fish: ⅓ to ½ pound per portion.
(3) Steaks and filets: ⅓ pound per portion.

Fresh whole or dressed fish should have these characteristics:

(1) Odor: fresh and mild.
(2) Eyes: bright, clear, and full.
(3) Gills: red and free from slime.
(4) Skin: iridescent.
(5) Flesh: firm and elastic, not separating from the bones.

In seafood, Grade A means top or best quality. Grade A products are uniform in size and relatively free from blemishes or defects. The products should have a good flavor.

Grade B means good quality, but the products may not be as uniform in size or as free from blemishes or defects. This is the commercial grade.

Grade C assures fairly good quality, but the same nutrition as the higher grade.

Frozen fish and shellfish should be frozen solidly, with very little or no odor. They should be packed in moisture-vapor-proof material with little or no airspace around the fish or seafood.

Oysters in the shell should be alive when we buy them and the shells should close up tightly when we tap them. Shucked oysters (out of the shell) should be plump and have a natural or creamy color and a small amount of clear liquid with a fresh, mild odor.

Clams in the shell should be alive when we buy them. Their shells should close tightly when they are tapped. Shucked clams should be pale to deep orange and have a fresh, mild odor with little or no liquid.

Crabs in the shell are sold fresh, frozen, or cooked. Fresh crabs should be alive when we buy them. Cooked crabs are bright red and have a mild odor. Softshell crabs are actually molting blue crabs after they have shed their shells. They should be alive and active when we buy them. Crab meat is removed from cooked crabs and should be white with an attractive red tint on the outside.

Fresh lobsters should be alive when we buy them. The tail of a live lobster will curl up under the body and should not hang down when the lobster is picked up. Lobster meat should be white with an attractive red tint on the outside.

Shucked scallops are the adductor muscles removed from the shells. The meat should be a creamy white, light tan, orange, or pink. Fresh scallops should have a sweetish odor and be packed in little or no liquid.

Raw shrimp range in color from greenish gray to brownish red, but when cooked, they all turn red. Cooked shrimp should have this attractive color and a mild odor.

STOCKING THE LARDER

POULTRY

We've come a long way from the days when chicken on Sunday meant wringing the fowl's neck, scalding, plucking, and singeing. Today, we can buy freshly dressed chickens, turkeys, geese, guineas, and ducks just as easily as we can buy a sack of flour and a bag of potatoes.

The number of servings we can get from different birds depends on the overall weight and how we prepare them. The lighter the bird for its species, the greater the proportion of the weight will be in the bone. The USDA general rule for whole poultry is to plan for ½ to ¾ pound per serving. This would mean we can get 5 servings from a 2½- to 3-pound broiler and 24 servings from a 12-pound turkey.

We can usually count on the following number of servings from these birds:

(1) Chickens
   Broilers: ¼ to ½ bird per serving.
   Fryers: ½ lb. per serving.
   Roasters: ½ lb. per serving.
   Stewers: ½ lb. per serving.
(2) Turkeys
   Under 12 lbs.: ½ lb. per serving.
   Over 12 lbs.: ½ lb. per serving.
(3) Ducks
   3–5 lbs.: 1 lb. per serving.
(4) Geese
   8–12 lbs.: ⅔ lb. per serving.

Federal law requires governmental inspection on all poultry shipped from one state to another. When we see the round USDA inspection mark on poultry and poultry products, we can be sure the product will be wholesome if it is handled properly.

Grade A is usually the only grade for poultry found in retail outlets. Birds graded below the U.S. Grade A are usually sold without the grade mark in supermarkets or are used in processed foods where appearance is not important.

Poultry is usually labeled according to age using these terms:

(1) Mature chicken: mature chickens, old chickens, hen, stewing chickens or fowl.
(2) Mature turkeys: mature turkey, yearling turkey, or old turkey.
(3) Mature ducks, geese, and guineas: mature or old.
(4) Young chickens: young chicken, Rock Cornish game hen, broiler, fryer, roaster, or capon.
(5) Young turkeys: young turkey, fryer roaster, young hen, or young tom.
(6) Young duck: duckling, young duckling, broiler duckling, fryer duckling, or roaster duckling (40).

We can almost always save money when we buy a chicken whole. Butchers make good salaries in supermarkets. When we pay for chickens which have been cut up, we're helping to pay those salaries. Not that we have anything against butchers, but why not save a few pennies and do the job ourselves?

EGGS

"Brown eggs are better for you." "Brown eggs are more nutritious." These myths have survived for years. Color makes no difference at all. It has no more significance than the fact some of us have red hair.

What *does* count, however, is the grade, the condition of the egg, and the price.

Size and quality are not related. They are entirely different. Large eggs may be high or low quality; high-quality eggs may be either large or small.

When we buy eggs, we should first look for the USDA grade shield on the carton. The grade will be shown within the shield. The size will be found on the carton.

There are 3 consumer grades: U.S. Grade AA (or fresh fancy), U.S. Grade A, and U.S. Grade B.

U.S. Grade AA and U.S. Grade A eggs have firm, round yolks and a high, thick white when they are broken out. Grade B eggs have thick whites and the yolks may be somewhat flatter than in a Grade AA or A (33).

The USDA sizes or weight classes are based on the minimum weight of eggs per dozen. These sizes and their minimum weights are:

Size	Weight per Dozen (USDA sizes in ounces)
Jumbo	30
Extra large	27
Large	24
Medium	21
Small	18
Peewee	15

Eggs are valuable as a protein food. When we consider the cost of eggs, we should remember that one dozen large eggs represent 1½ pounds (24 ounces) of high-quality protein in the shell. If the large eggs are selling for 90 cents a dozen, that's the equivalent of 60 cents per pound, a reasonable price for a pound of protein-rich food.

We should never buy dirty or cracked eggs, because they may contain bacteria that cause food illnesses.

In many supermarkets, clerks check eggs to make sure they are whole and sound before the customer takes them home. If we happen to crack an egg accidentally before we are ready to use it, we don't have to throw it out. Cracked or soiled eggs should be used only in foods that will be thoroughly cooked before eating, such as baked goods or casseroles.

We should make sure we buy eggs which are displayed in a refrigerated display case and refrigerate them promptly at home to help maintain quality.

CHEESE

There was a time, not too many years ago, when the only cheese most Americans ate was in macaroni and cheese or in pimento cheese sandwiches. Now that has all changed, because Americans have become some of the most enthusiastic cheese eaters in the world.

The best way to buy cheese is in the bulk. A whole cheese can be inspected (and maybe even sampled). However, most supermarkets do not sell it in bulk. Cheese stores or large stores with cheese departments are more likely to offer it this way.

The way a store keeps its stock of cheese is often a tip-off to the quality of the cheese itself. If the store is clean and the cheese is kept covered and refrigerated and the store has a brisk business, the cheese should have a rapid turnover and be in good condition.

However, most of us buy our cheese in supermarkets where it is sold in smaller, prewrapped packages.

We should check the packages carefully to make sure the wrappers are clean and not torn or stained. The cheese should be kept under refrigeration.

Labels for processed cheese must list all ingredients used in the preparation of the cheese, and tell us if the cheese we are buying is processed, cheese food, or cheese spread.

It is better to buy small quantities of the soft, unripened natural cheeses such as cottage or cream cheese, and the soft, ripened cheeses such as Camembert and Brie. They are perishable, and they lose their flavor and texture in a few weeks. Firm, ripened cheeses and processed cheese will keep far better, so we may safely buy them in large quantities (16).

Cheese grades usually do not appear on retail packages, but occasionally we may find the U.S. Grade AA mark on some packs of Cheddar cheese. Cheeses of this quality are excellent with a fine, pleasant flavor. Grade A is also a very good quality.

Processed cheese and cheese foods do not have official grades; however, they may be officially inspected and bear the USDA "Quality Approved" inspection shield. Cottage cheese may also bear this shield. To earn the shield, cheese must meet high quality standards and must have been manufactured in a plant which meets USDA standards for plant and equipment conditions.

The best way to find the relative cost is to compare the price of equal weights of cheese.

STOCKING THE LARDER

FRESH FRUIT

The USDA provides inspection service for certification of the quality of fresh and processed fruits, based on the U.S. grade standards. This is strictly a voluntary service offered to the fruit industry on a fee basis. The service is widely used.

Often we find prepackaged apples and some citrus fruits will bear a grade, but fruit in the bulk seldom does. The top grade in most cases will be either U.S. Fancy or U.S. No. 1. Both are excellent grades (34).

To find the perfect peach, the perfect watermelon, both at their peak, is not an easy task.

Appearance and quality are related in many respects, but a good appearance does not necessarily mean good quality. Often a fruit with a very attractive appearance may have relatively poor eating quality, because of some internal conditions such as overmaturity. On the other hand, a fruit with relatively poor appearance due to superficial blemishes may have excellent eating quality. However, poor color in normally well-colored fruit may indicate inferior flavor.

Of course, the best time to buy fruit is when it is in season and can be bought for less. It is wasteful to buy more bananas or oranges than our families can eat, just because they happen to be on sale that week. Unless the lower price is the result of in-season overabundance, our bargains may turn out to be bad buys. Modern refrigeration makes it possible to keep some fruit in good condition for a few days, but we cannot fill a refrigerator with fruit. There must be room for other foods as well (and maybe even the ironing).

The best place to buy fruit is in a store which has a fairly rapid turnover in fresh produce and one which allows us to choose our own fruit. We should watch closely for any signs of deterioration. Even though most stores take considerable pains to keep their fresh fruit in top condition, some deterioration will take place in display bins. Frequently, we can buy off-quality fruit at reduced prices but, when we have to cut off soft, decayed, or damaged portions, price reduction is offset by the waste.

Large-sized fruits are not necessarily the best or the most economical. They may look like bargains, but they may be mealy or completely unsuited to our needs. From a quality standpoint, however, it is better to buy large fruit than fruit that is too small.

Even though we must handle produce when selecting it, we should pinch ourselves to remind us not to pinch fruits and vegetables. Rough handling causes more rapid deterioration.

Apples

They should be firm, crisp, and well-colored for the variety. We should avoid overripe apples (indicated by a yielding to slight pressure on the skin and soft, mealy flesh) and apples which are badly bruised or affected by decay. Scald on apples (irregular-shaped tan or brown areas) may not necessarily affect the eating quality of an apple, but does indicate the apple has been in storage for a long period.

Apricots

Apricots should be plump and juicy-looking with a uniform, golden orange color. Ripe apricots will yield to gentle pressure on the skin. We should avoid soft or mushy fruit and very firm, pale yellow or greenish yellow fruit.

Bananas

Firmness, bright appearance, and absence of bruises and other injury are characteristics of good bananas. They should also be solid yellow, flecked with brown, for immediate use. Yellow bananas with greenish tips will ripen at home in a few days.

Blueberries

Blueberries should be dark blue with a silvery bloom, and be plump, firm, dry, and free from stems or leaves.

Cherries

Sweet cherries should have a very dark color. We should avoid cherries which have dried stems and a dull appearance. Decay is fairly common at times on sweet cherries, but because of the dark color, decayed spots are inconspicuous. Soft, leaking flesh, brown discoloration, and mold growth are all signs of decay.

Cranberries

Plump, firm cranberries with a lustrous color are best.

Grapefruit

We should choose firm, well-shaped fruits which are heavy for their size. Thin-skinned fruits have more juice. Rough, ridged, or wrinkled skin can be an indication of thick skin, pulpiness, and lack of juice. Grapefruit often have skin defects such as scale, scars, thorn scratches, or discoloration; these usually do not affect the eating quality.

Grapes

Grapes are at their peak when they are well-colored, plump, and firmly attached to the stems. Avoid soft or wrinkled grapes, those which have bleached areas around the stem end, or which are decayed, wet, or leaking.

Lemons

Lemons should have a rich, yellow color, be reasonably smooth with a slight gloss, and be firm and heavy.

Limes

The best limes have glossy skins and are heavy for their size. Purplish or brownish irregular mottling of the outer surface is a condition called "scald" and does not damage the flesh.

Melons

The selection of melons for quality and flavor can even be difficult for experts. Here are a few guidelines which will increase the odds in our favor (34):

(1) Cantaloupe: the stem should be gone, leaving a "full slip." If all or part of the stem base remains, or if the scar is jagged or torn, the melon is probably not fully matured. We must also look for netting or veining that is thick, coarse, and corky. It should stand out in bold relief over some part of the surface. The skin color (between the netting) should have changed from a greenish color to a yellowish buff, yellowish gray, or a pale yellow.

(2) Casaba: ripe melons will have a gold-yellow rind color with a slight softening at the blossom end. Casabas have no odor or aroma.

(3) Crenshaw: ripe melons should be a deep golden yellow, sometimes with small areas having a lighter shade of

yellow. The surface should yield slightly to moderate pressure of the thumb, particularly at the blossom end. Ripe melons should have a pleasant aroma.

(4) Honey Dew: ripeness will be indicated by a soft, velvety feel. There will be a slight softening at the blossom end, a faint pleasant fruit smell, and a yellowish white to creamy rind color. Avoid melons with a dead white or greenish white color and a hard, smooth feel.

(5) Watermelon: even the experts make mistakes. The safest way to buy watermelon is to buy a cut melon which has juicy flesh with a good red or gold color, free from white streaks, and which has brown or black seeds. When we buy an uncut melon, we should look for relatively smooth surfaces, rinds which are slightly dull (never shiny), ends which are filled out and rounded and a creamy colored underside or "belly."

Nectarines

Rich color and plumpness as well as a slight softening along the "seam" of a nectarine indicate ripeness. Avoid hard, dull fruits or slightly shriveled ones.

Oranges

Firm and heavy oranges with fresh, bright-looking skin which is reasonably smooth indicate good-quality oranges. Avoid lightweight oranges, which are likely to lack flesh and juice. Very rough skins mean abnormally thick skin and less juice.

Peaches

They should be fairly firm and should be starting to get a little bit soft. The skin color between the red areas should be yellow, or at least creamy. Avoid very firm or hard peaches with distinctly green background color. They will probably not ripen correctly. Avoid very soft, overripe fruits. We should never buy peaches with large, flattened bruises. Decay starts as a pale, tan spot and enlarges in a circle, gradually turning darker.

Pears

Pears of all varieties should be firm. The color depends on the variety. Pears which are hard when we buy them will probably ripen if we keep them at room temperature, but it is better if we select fruit which have already begun to soften. Avoid pears with dull skin which are wilted or shriveled with weakening of the skin near the stem, or with sunken spots on the sides or blossom ends.

Pineapples

Color and size alone are not always dependable guides. A yellow rind is not necessarily an indication of a good ripe pineapple. Many pineapples in prime condition have what is known as a "chocolate-green" color, or mottled green and brown. Best quality fruit will usually have a small compact crown on top. Pulling leaves from the crown is not a dependable test for ripeness. The best guide is to snap the side of the fruit with the thumb and finger. A dull, solid sound indicates a well-ripened, sound fruit, full of juice. A hollow thud means the fruit is sour, not well-matured, and lacking in juice.

Plums and Prunes

We should choose plums and prunes with a good color for the variety and which are fairly firm to slightly soft.

Raspberries, Boysenberries, etc.

Berries should be bright, have a clean appearance and a uniform good color for the species. The individual small cells making up the berry should be plump and tender, but not mushy. There should be no stem caps attached.

Strawberries

Ripe strawberries are a full, red color and have a bright luster, firm flesh, and have the cap stems still attached. The berries should be dry and clean. Small to medium-size berries usually have a better flavor than larger ones.

Tangerines

Ripe tangerines have a deep yellow or orange color and a bright luster. Because of the typically loose nature of tangerine skin, they will frequently not feel firm to the touch.

FRESH VEGETABLES

Fresh vegetables are inspected and graded extensively by packers, processors, buyers, and others in wholesale trading as a basis for establishing the value of a product. This insures a minimum quality in the supermarket.

Sometimes potatoes, onions, carrots, apples, and some citrus fruits in retail stores are packaged in containers which give grade designations. But this does not necessarily mean the product has been officially graded unless the product also bears the official USDA grade shield or the statement "Packed Under Continuous Inspection of the U. S. Department of Agriculture" or "USDA Inspected."

We can usually find the best-tasting and lowest-priced vegetables if we buy when they are in season. As a general rule, young vegetables are the tenderest and best tasting. A big vegetable is not necessarily the best vegetable.

Freshness is important to both flavor and quality, so we should never buy more vegetables at one time than we can consume or preserve within a week.

Artichokes

We should avoid artichokes with large areas of brown on the scales and with badly spreading scales. (This is a sign of age.) Avoid decay, mold growth, worm injury, or grayish, blackish discoloration.

Asparagus

Asparagus spears should have tips which are compact. Spears with decayed tips, and ribbed spears with up and down ridges should be avoided. When the lower portion or butt end is white, the spear may be excessively fibrous and tough.

Snap beans

They should be fresh-looking with a bright, crisp appearance and good color.

Beets

Fresh beets should be firm and round with a slender taproot, and have a rich, deep-red color with a relatively smooth surface. If beets are bunched, we should avoid badly wilted tops.

Broccoli

Broccoli of good eating quality should have firm and compact clusters of small flower buds, but with none opened enough to show the yellow flowers inside. The color should be green and may have a purplish cast. The stems should not be too thick or tough.

STOCKING THE LARDER

Brussels Sprouts

Sprouts should have a fresh, bright-green color with tight-fitting outer leaves and firm bodies. Avoid sprouts with yellow, ragged, or worm-eaten leaves.

Cabbage

Good-quality cabbage should be firm to hard, and the heads should be heavy for their size. In new cabbage, outer leaves should have a good green or red color. Cabbage that has been stored may be closely trimmed down to the white inner leaves.

Carrots

They should be well-formed, smooth, well-colored, and firm.

Cauliflower

Heads with white to creamy white, compact, solid, and clean curds are best. A white granular or "ricey" texture of the curds will not hurt the eating quality if the surface is compact. We should avoid cauliflower which has spreading curds or shows mold or decay.

Celery

Freshness and crispness are all-important. Stalks should be glossy and light to medium green and have a solid, rigid feel; the leaves should be fresh or only slightly wilted.

Corn

Corn should have succulent-looking husks with good green color; the silk ends should be free from decay or worm injury, with the stem ends not too discolored or dried. Ears should be well-covered with plump, milky kernels.

Cucumbers

Those which have a good, green color and are firm are best. Good-quality cukes should not be too large or overgrown and dull.

Eggplant

A good-quality eggplant should be firm, heavy for its size, smooth, and uniformly dark purple.

Greens

Leaves should be fresh, young, tender, free from blemishes, and have a healthy, green color. Beet tops and ruby chard will have a reddish color.

Lettuce

Iceberg lettuce heads should be firm but not too hard, fresh and crisp. Romaine lettuce should also have crisp leaves. Other varieties of lettuce will have softer textures. Outer leaves should not be wilted; color should be bright.

Mushrooms

Caps should be either closed around the stem or only slightly open, and gills should be pink or light tan. The surface of the caps should be white or creamy. However, mushrooms from some producing areas have light brown caps. Young and tender mushrooms which are small to medium in size are best.

Okra

Tender pods with tips which will bend with slight pressure are more desirable. Always avoid tough, fibrous pods.

Onions

The best-quality onions are hard, firm, and dry and have small necks. They should be covered with papery outer scales and be reasonably free from green sunburn spots and other blemishes.

Green Onions

Tops should be fresh, crisp, and green.

Parsnips

The best parsnips are small to medium-sized and are well-formed, smooth, firm, and free from serious blemishes or decay.

Peppers

Good peppers are relatively heavy for their size and have firm walls or sides. They should be medium to dark green color with a glossy sheen.

Potatoes

New potatoes should be firm, well-shaped, and free from blemishes and sunburn (a green discoloration). Some amount of skinned surface is normal, but potatoes with large skinned and discolored areas are undesirable. General purpose and baking potatoes should be smooth, well-shaped, firm, and free from blemishes, sunburn, and decay. We should avoid potatoes with large cuts or bruises, those with a green color (sunburn) and any showing signs of decay or those which are badly sprouted or shriveled.

STOCKING THE LARDER

Radishes
We should look for medium-sized radishes which are plump, round, and firm.

Rhubarb
Good rhubarb stems have a large amount of pink or red color, and are glossy, bright, fresh, and firm. Avoid rhubarb which is wilted and flabby, extremely slender or extremely thick.

Squash (Summer)
Good summer squash should be tender and well-developed, firm, and fresh-appearing. A tender squash will have a glossy skin which will not be dull or tough.

Squash (Winter)
Fully matured winter squash will have a hard, tough rind. We should buy winter squash which are heavy for their size and which have only slight variations in skin color. Squash that are cut, punctured, or have sunken or moldy spots should be avoided.

Sweet Potatoes
We should look for well-shaped, firm sweet potatoes with smooth, bright, uniformly colored skins and which are free from signs of decay. Since they are more perishable than Irish potatoes, we should use extra care in selecting them.

Tomatoes
Tomatoes that are of good quality will be well-formed, smooth, well-ripened, and be reasonably free from blemishes. Avoid overripe or bruised tomatoes, or those with yellow areas near the stem scar or which are badly growth-cracked. Tomatoes which have soft, water-soaked spots, depressed areas, or surface mold have begun to spoil.

Turnips
The best white turnips are small to medium in size, and are smooth, fairly round, and firm. If they are sold in bunches, the tops should be fresh.

STAPLES

We reach for some of our staple items every time we prepare a meal. They are the backbone of our cooking efforts. If we select them with care, they will do a lot to establish our reputations as

KEEPING FOOD SAFE

wise consumers and good cooks. We should always be certain a supply of any staple is used up before we add more to the container. Otherwise, we may just keep adding sugar, tea, or coffee on top of old products.

CANNED PRODUCE

We need to train ourselves to be label readers. Manufacturers are required by law to list everything used in the preparation of canned products, except for certain "standardized" items. The fair-packaging and labeling regulations enable us to take a quick look at the label and tell exactly what we are buying. Labels may also make it easier for us to compare prices.

U.S. Grade A or Fancy canned produce is the tenderest, most succulent, and flavorful. U.S. Grade B or Extra Standard produce is also of excellent quality, but not necessarily quite as uniformly colored. They are not as tender as Grade A. U.S. Grade C or Standard produce is not as uniform in color and flavor and is usually more mature. However, it is a thrifty buy when appearance is not important.

Never, never under any circumstances, should we buy cans which are bulging or which have an imperfect seal.

FROZEN FOODS

The freezer in our supermarket is just as important as the one in our home. It should be clean and not overloaded. Most freezing cabinets in stores have a line around the inside indicating the top limit for storing packages. We should make sure the food we buy is not stored above this line. Packages of frozen food should be solid to the touch with no stains or punctures.

PART 2

Refrigeration and Freezing

If we were to step into the twentieth century from the late 1800s, the cultural shock we would experience would be the sharpest right in our own homes.

The television set in the family room, for instance, has changed society irrevocably. The pushbutton kitchen spawned the women's movement. Refrigerators and freezers have caused revolutions in our buying and eating habits.

REFRIGERATION

Countless cases of food illness could be prevented each year if only all of us understood how very critical cool temperatures are in controlling the growth and multiplication of microorganisms.

The pioneer homemaker in the nineteenth century was constantly concerned about temperature control, because she had no easy way to keep foods cool. We have grown lax, because refrigeration is so easy, so simple.

Which kinds of food require refrigeration? All perishable products, including potentially hazardous foods must be refrigerated. Those foods we class as potentially hazardous are animal products or those which contain at least some animal products. They are milk, milk products, eggs, meat, poultry, fish, and shellfish. Also included are cream pies, custards, potato salad, and meat salads.

Although these foods may look fine and taste delicious when we eat them, sometimes we cannot tell until it is too late if they are contaminated.

These potentially hazardous foods must be refrigerated immediately after we bring them home from the market. (We should come straight home from the market.) They must be kept refrigerated except during the time actually needed to prepare and serve them.

Leftovers containing even small portions of hazardous foods must never be left at room temperature after a meal, but refrigerated immediately.

Of course, these potentially hazardous foods will not remain safe for us to eat if our refrigerators do not keep them cold enough. Foods should be stored in small, shallow containers. The more surface exposed to the cold, the faster food will cool. Containers must be kept covered so food particles from the shelves above will not fall into dishes below.

Food must not be crowded into our refrigerators, because in

KEEPING FOOD SAFE

order to protect foods, cold air must circulate freely around and between them.

Frequently, we keep many things in our refrigerators which do not need to be refrigerated. We need to allow plenty of space for perishable foods and leftovers to chill properly. Condiments such as mustard and catsup do not need to be kept chilled, and soft drinks can be chilled just before serving.

It's a good idea to keep leftovers in one spot in the refrigerator so they will be in plain view, and we will remember to use them while they are still fresh.

Above 45° F. (and up to 115° F.) both infectious bacteria and toxin-producing microorganisms may grow rapidly. This is why it is critically important that we prepare and serve perishable foods with as little time lapse as possible after taking them from the refrigerator or freezer. After a meal is over, putting away the leftovers should be the first order of business.

Some foods can be kept in the refrigerator for several weeks, but since refrigeration does not kill bacteria (only retards their growth) we should make every effort to use perishable foods within a reasonably short time.

Our refrigerators have revolutionized our dietary habits. They have made it possible to keep some foods very cold, yet others only moderately cool. Since refrigerators are constructed to take advantage of the principles of normal movements of cold and hot air, it's a good idea to know just how cold the various sections are.

Temperatures will vary in any refrigerator, depending on how it is constructed and used. We should use a thermometer made especially for the purpose to take its temperature. Generally, the air in a refrigerator is coldest next to the freezer and warmer farther away from it. Frost-free models usually have less variation in temperature. Table 8 shows where to store food to take advantage of a refrigerator's temperature variations.

Since air is almost constantly circulating in a refrigerator, food tends to dry out easily. We should cover most foods.

The lengths of time foods can be stored under refrigeration vary. (Part 3 of this chapter discusses food storage in detail.)

With all of the help appliance manufacturers give us, there

should be no excuse for anyone to become ill from a foodborne disease caused by improper refrigeration.

However, until now there has been no single easily accessible source of information on proper storage measures for all foods. When refrigerators originally came into our kitchens, the first reaction was relief, because consumers no longer had to rely on the iceman.

But we've gone off on a tangent as far as our refrigerators are concerned, and we have forgotten some of the elementary rules for safeguarding food.

However, millions of dollars are now being spent on a consumer education program aimed at stamping out foodborne disease in the United States. Much of this effort is aimed at teaching us how important refrigeration is.

We have been incredibly trusting in the past few decades. We have believed we were almost immune to foodborne illnesses because we had our glossy and sleek appliances.

Today's manufacturers build sophisticated and highly efficient appliances, but they are no better than the decisions we make when we use them. We should keep seven fundamental rules in mind when we consider refrigeration:

(1) We should buy and use an accurate refrigerator thermometer.
(2) We must keep the inside of our refrigerators clean. They should be washed regularly.
(3) The areas around the motor and refrigerating unit must be kept free of lint and dust to keep air circulating evenly and to keep the machinery operating efficiently.
(4) We should check the gaskets around refrigerator doors fairly often. They must be flexible in order to hold the cold air in.
(5) In a conventional refrigerator, ice must not be allowed to build up more than ¼ inch on the cooling coils. The frost will act as an insulator, and the refrigerator will have to work harder to cool.
(6) Food should be stored in small, shallow containers. The

more surface exposed to the cold, the faster the food will cool. Foods should be kept covered so food particles from higher shelves will not fall into food stored below them.

(7) We must make certain there is ample space between foods and between the walls of the refrigerator. There must be free circulation of air. This is why we should never cover the wire shelves of a refrigerator with paper or foil. Foods should never be stacked in a refrigerator.

A large percentage of the foodborne diseases which occur each year are a result of failure to keep potentially hazardous foods cold. We can minimize our chances of getting foodborne illnesses if we follow those seven simple steps.

The all-important universal food safety rule is to either keep foods cold (40° F. or lower) or keep them hot (115° F. or above), or just don't keep them.

FREEZING

Although refrigeration produced a revolution in our dietary habits, freezing has made almost as many changes in the way we buy and store foods. Freezing drastically reduces the spoilage in food that naturally occurs from bacteria, enzyme action, and oxidation. When we freeze foods we are able to maintain most of the quality, texture, and flavor of fresh food.

The Associated Food and Drug officials of the United States have recommended that all frozen foods be held at 0° F. or lower, although brief periods of warming up to 10° F. can be tolerated for either loading or unloading or during a temporary power failure. (Fluctuating levels of coldness also tend to dry out all foods and make them more susceptible to the development of off odors.)

Microbial growth is not likely to occur at 10° F. In fact, there is a slow but continuous decrease in the numbers of microorganisms as freezer storage continues. Yet molds can grow in frozen foods at 16° to 20° F. or even lower, and yeasts can grow at 16° F. and above.

Foods tend to lose moisture easily in a freezer because of the frequent circulation of cold air. When this happens, foods become dry and flavorless. It is essential that we keep our frozen food in sturdy, moisture-vapor-proof packaging such as rigid or flexible plastic, heavy-duty aluminum foil, wax-coated freezer paper, heavy-duty plastic freezer bags or plastic-coated cardboard containers which will seal tightly. We should never use non-rigid containers which are not moisture-vapor-proof. (This rules out the plastic bags which bread comes in, ordinary cardboard boxes, and paper sacks.) Packages should be sealed completely; we should never fasten a box or sack with plastic tape, because it cannot hold a seal in the freezer.

As food supplies dwindle, more and more of us are tempted to pack our freezers to the brim whenever we see a bargain. If we pack them with too much food, improper temperatures can cause loss of nutrition, flavor, texture, and protein. There can also be dangerous microbial growth.

We should freeze only as much food as our freezers can handle at one time without raising the temperature above 0° F. The U. S. Department of Agriculture recommends that no more than 2 pounds for each cubic foot of freezer space be frozen every 6 hours (25).

The Department also recommends that we freeze foods as quickly as possible. In order to do this, we must place packages so they will contact a surface in the freezer. There must also be space all around for air to circulate. After the packages have been frozen, we can stack them together.

Although many foods can be frozen very satisfactorily, many should not be frozen:

> Pie meringue, egg white frosting, custard pies, layer cakes with soft fillings, sour cream, buttermilk, yogurt, creamed cottage cheese, eggs in the shell, hard-cooked eggs, potato salad, fully cooked fried foods, cracker canapes, completely cooked spaghetti, macaroni, noodles, rice, mayonnaise, lettuce, celery, radishes, oil-based salad dressings, unblanched vegetables, raw potatoes, gelatins, some refrigerated doughs, bacon, and seasoned sausages.

KEEPING FOOD SAFE

No freezer should ever be placed near a furnace. Freezers should stand where the temperature will range somewhere between 50° and 70° F.

No freezer can operate efficiently when there is a big frost buildup. There are no set rules about how often a deep freeze should be defrosted, except that when frost is ½ inch thick, we should remove it. The number of times we open the freezer door and how well we have wrapped the food are important factors in determining how quickly frost builds up.

When we defrost our freezers, we should first unplug them, then quickly transfer the frozen food to picnic iceboxes or pack it in newspaper-lined boxes and cover with blankets. The interior of the freezer should be washed with a soap, water and soda solution, then rinsed well and dried with a terry cloth towel. It's also a good idea to air dry the freezer for a few minutes before the current is turned back on.

One of the most frustrating experiences we can have with a freezer is trying to decide what to do when the power goes off. After all, even a 25 percent rise in temperature over a full day's time can hasten spoilage and destroy nutrients.

If we suspect a power outage may be lengthy, it's a good idea to buy some dry ice. (Dry ice has a temperature of $-110°$ F.) A 50-pound block of dry ice will keep our foods frozen safely for 2 to 3 days, depending upon how full the freezer was when the power went off. A full freezer will maintain safe, low temperatures longer. (Of course, we should not open the freezer door unnecessarily once the dry ice has been put in place.) If we have to resort to dry ice, there are certain rules we should follow:

(1) Always wear heavy gloves when handling dry ice.
(2) Never try to chip dry ice with an ice pick, because the splinters are like slivers of glass and they can put out an eye.
(3) Do not open the freezer door any more than necessary.

In a 4- to 6-cubic-foot freezer, food will not thaw if dry ice is being used for the first 15 to 20 hours. However, after about 48

STOCKING THE LARDER

hours, the temperature will go up to about 40° F., the normal refrigerator temperature.

Foods subjected to this treatment may be refrozen if they still contain ice crystals. However, we must not forget that low-acid foods like vegetables, meats, poultry, and fish which have reached a temperature of 50° F. must be cooked immediately or thrown out.

ICE-CUBE COMPARTMENT FREEZING

As we have learned, all frozen foods should be held at 0° F. or lower, although brief periods of exposure up to 10° F. can be tolerated for defrosting, loading or unloading, or during a power failure.

The ice-cube compartments of ordinary one-door refrigerators do not hold frozen goods at that critical 0° F. point. They were not designed to store frozen foods. An ice-cube compartment will keep ice cubes frozen very nicely at any temperature below 20° F., but we have learned that bacteria starts growing at 10° F. The temperature in a compartment may well read as high as 15° to 20° F. We have also learned that frozen foods begin to thaw in this temperature range, even though the packages may feel hard. One day at 20° to 25° F. does more damage to foods than a full year in the freezer at 0° F.

If we must store frozen foods in the frozen food compartment of a conventional, one-door refrigerator, we should plan to use them within a period of 1 week. The American Frozen Food Institute says foods frozen in the ice-cube compartment of a conventional, one-door refrigerator can be frozen for several days only. The Institute recently tested ice cream, orange juice, and strawberries in an ice-cube compartment, and they found the quality had deteriorated in about 7 days.

There is really no way to get around this problem, because if the controls in the refrigerator are set low enough to bring the ice-cube compartment down below 15° F., other food items in the refrigerator will also freeze.

Deterioration in frozen foods accelerates rapidly with a rise in temperature.

For example, a group of experienced food tasters were able to detect a change in flavor from the original freshness of frozen strawberries after a year's storage in 0° F., after 2 months' storage at 10° F., and after only 9 days at 30° F. (18).

If we have to use our ice-cube compartments as storage space for frozen foods, there are 5 rules we must follow carefully:

(1) Keep the temperature control at the setting just above freezing in the general storage area.
(2) Buy frozen food just before checking out at the store.
(3) Buy only enough food to be used within one week.
(4) Ask the grocery packer to put frozen foods in an insulated bag or a double-paper bag, then take the foods home, take them out of the bag, and put them in the ice-cube compartment as quickly as possible.
(5) Frozen food should be placed in contact with the floor of the ice-cube compartment or the refrigerated ice tray shelf.

Home storage in an ice-cube compartment should not be an endurance test for frozen foods. Planning for the use of frozen foods within the framework of the capabilities of our home equipment is the key to maintaining high-quality food during storage.

Any time we use either an ice-cube compartment or a two-door refrigerator-freezer, we should check the temperature occasionally with a thermometer. We can buy an inexpensive freezer-refrigerator thermometer at almost any supermarket or hardware store. The thermometer should be placed in front of the storage area, up high in the food load. We should leave it there at least overnight without opening the freezer. If the reading is above 0° F., we should adjust the controls in order to bring the temperature down. Then we should wait for another overnight period when the freezer will not be opened and check it again. We need to make a habit of reading the thermometer anytime we open the freezer door. If we have a frost-free freezer, we should read the thermometer when the fan is going, not when the appliance is in its automatic defrosting cycle.

STOCKING THE LARDER

We cannot remind ourselves too often that hard-frozen packages are not in themselves proof that the temperature is cold enough to protect the food at all times.

When we bring frozen food home from the market, it should be placed in the bottom or the back of the freezer so we will be sure to use the oldest food first.

Packages should be marked with the contents and date of freezing. It also helps to keep a list of items beside the freezer. Items can be checked off as they are used.

PART 3

Storing

Food is becoming more precious with each passing year. It is imperative that we take good care of what we have.

Every family should have a 2-week supply of foodstuffs on hand to see them through any type of emergency. Yet a consumer who stores food improperly is running more than one risk.

Proper storage of food means putting it away promptly under the best conditions and keeping it away from filth, pests, and household chemicals.

We have already learned that frozen foods must be kept in a freezer at 0° F. or lower for long-term storage, and that perishables need to be refrigerated at 40° F. or lower.

We must be careful and use our common sense when we store foods, because perishable items are sometimes packaged in plastic bags, jars or cans that make them resemble nonperishables.

If a perishable is accidentally stored without refrigeration, it may well become unsafe to eat, because of bacterial growth. If this happens, it should be discarded even if it looks or smells good.

A cabinet with a drainpipe running through it (such as the area under a kitchen sink) is not a safe place to store food, because there may be leakage, and because it is almost impossible to completely seal up the openings the pipes pass through. Rodents and insects can easily enter this type of area.

KEEPING FOOD SAFE

We need to give serious thought to maintaining high standards in our storage areas. Cupboards should be cleaned frequently. Dust can contaminate food with microorganisms when stored cans or packages are opened. Can lids should be wiped with a sudsy cloth before we open them.

Never, under any circumstances, should we keep household chemicals in food storage cabinets. Cleansers, solvents, paints, polishes, pesticides, and cosmetics have found their way into foodstuffs when a consumer mistakenly thought they were ingredients.

Some sort of identification system is invaluable in food storage areas, so we will be sure to use oldest items first. Most foods deteriorate, even under the best of circumstances. When they are kept too long, most of them will deteriorate in quality. If the container deteriorates too, the food may well become contaminated.

Anytime we prepare foods which have been in storage for a long period, we need to examine them carefully for abnormal appearance or odor which might indicate spoilage. Of course, under no circumstances should we ever taste any food which we suspect may be spoiled. These foods should be completely destroyed so not even animals can consume them.

Since extra foods are often kept in basements, some of us may have flood damage once in a while. We must keep in mind that floodwaters carry disease organisms. Even rain water that pours in through a basement window will have some pathogens in it.

Fresh fruits and vegetables contaminated by these waters should be destroyed or peeled and cooked immediately. Any cardboard boxes containing cereals or bottles of flavoring should be destroyed. We should also throw out any capped bottle, such as those containing catsup and soft drinks.

Foods preserved in cans may become infected if seals are broken. These foods should be discarded.

Even if a sealed can has not been punctured during the flooding, special precautions must be taken before we can use these products. The cans must be boiled in water which covers them for at least 15 minutes. The labels may come off during the process, so the cans should be remarked. Food packed in glass jars should be discarded.

STOCKING THE LARDER

MEAT

When we bring meat home from the market, we should remove or loosen the commercial wrapping around it unless we plan to cook the meat right away. Washing meat before storing it is a no-no, because the added moisture may well encourage the growth of spoilage organisms. However, we can wipe it with a damp cloth.

Chopped and ground meat, and the variety meats like brain, heart, liver, etc., are all extremely perishable and should not be refrigerated for more than 1 or 2 days. We should cook or freeze variety cuts which we cannot use soon after purchase.

The cured meats (ham, fresh sausages, frankfurters, and luncheon meats) should be refrigerated in their original wrappings, and after opening they should be sealed up tightly. They should not be kept more than 7 days in the refrigerator.

Freezing is not recommended for cured meats, because the seasonings added for curing accelerate the development of rancidity. If we must freeze cured meats, they should be used within a month or two.

We have learned it is essential to cover and refrigerate leftover meats immediately after a meal. Leftover and unfrozen meat is best if used within 2 days. Cooked meat and many combination dishes can be frozen for 2 to 3 months.

Fresh or cooked meat which we do not plan to cook and use within the recommended storage time should be tightly wrapped and frozen.

Many small (1½- to 3-pound) canned hams do not need to be refrigerated. If they are not labeled "Perishable, Keep Under Refrigeration," or if they are dry-cured hams (country style and Smithfield), we may safely keep them in a cool, dry place until we use them. All other hams must be refrigerated.

For best quality, a cured whole ham should be kept in the refrigerator no longer than 7 days. Any cuts of cured ham smaller than a whole ham will keep safely in the refrigerator for 3 to 5 days.

Unopened canned hams labeled "Perishable, Keep Under Refrigeration" can be kept safely in the refrigerator up to 6 months, without loss of quality.

According to 1971 statistics on foodborne illness, ham was the meat most often involved in cases of "Staph" food poisoning. The "Staph" organisms involved in these illnesses usually came from humans.

Some hams come packed in plastic cans. Any unused ham can be safely stored in such a container.

All canned meats, except those marked "perishable," should keep well in a cool, dry place. However, they should be used within a few days once they are opened. Any unused portion may be left covered in the can and refrigerated.

Canned meats often contain storage instructions on the labels, and these should always be followed carefully. In general, however, canned meat should be protected from heat, from dampness, and from freezing temperatures. Canned meats should never be stored in direct sunlight, near hot pipes, or near a furnace.

Freezing does not cause canned meat to spoil, but it may damage the seal of the can so that spoilage begins. Dampness may also corrode cans or metal jar lids which in turn may let spoilage organisms enter.

Fresh poultry should never be stored in the refrigerator for more than 3 days. It should be loosely wrapped. The circulation of air around it helps keep down the growth of bacteria. Poultry should be stored in the very coldest part of the refrigerator. Before we store it (if we do not intend to cook it the same day we purchase it), we should remove the plastic film wrapper and take out the neck and giblets. Frozen poultry keeps best at 0° F. or lower and must be packaged in moisture-vapor-proof materials.

If we get a beautiful, smoked turkey at Christmas, we can keep it well-wrapped in the refrigerator until New Year's. If it must be kept longer than 2 weeks, it must be wrapped in a vapor-moisture-proof wrapper and frozen.

Canned poultry should be kept in a dry, cool storage area at temperatures below 70° F. for no longer than a year.

STOCKING THE LARDER

FISH AND SEAFOOD

It goes without saying that we are going to do our best to buy the freshest fish we can find. Now we know we should cook it immediately if possible.

If we have to keep fish or seafood for a short time before cooking it, we should wrap it in heavy wax paper and store in the meat keeper.

Fish and seafood should be handled gently. Bruised or punctured flesh will deteriorate more rapidly.

Prepacked fish and shellfish can be stored in the refrigerator in their original packaging. These wrappings or containers are designed for short-time refrigeration. Fresh fish and shellfish wrapped in butcher paper should be unwrapped, placed on a platter or tray, then covered with aluminum foil or plastic wrap before refrigerating. No fish or seafood must ever be held in the refrigerator for more than 2 days before cooking.

We must protect any fish or seafood to be frozen by wrapping or packaging carefully in moisture-vapor-proof materials. The kind of packaging should be suited to the kind, shape, size, and consistency of the food. Frozen fish and seafood must be stored at 0° F. or below, and should not be stored longer than 6 months.

Cooked fishery products may be stored either in the refrigerator or the freezer. If we store them in the refrigerator, cooked fish or shellfish should be kept in a covered container. We should never hold cooked seafood in the refrigerator for more than 2 or 3 days. For the best-eating quality, cooked fish or shellfish should not be stored in the freezer longer than 2 or 3 months.

Refrigerator thawing is safest for fish. (Allow 24 hours for a 1-pound package of fish to thaw.) We should never, never thaw fish or shellfish at room temperature or by immersion in warm water. Frozen breaded fishery products should not be thawed before cooking.

Canned tuna fish and salmon we have stashed away should be stored in a cool, dry place and we should use it within a year.

EGGS

Eggs should be refrigerated, large end up, to help maintain good quality. Ideally, eggs should be stored in their carton or some other closed container in the refrigerator. Research has shown eggs lose flavor and moisture if they are left uncovered, because the shells are porous.

At refrigerator temperatures, shell eggs will keep for several weeks, but we should use them within 1 week if we want to get best flavor and maximum quality.

Leftover yolks should be refrigerated promptly in closed containers. They should be covered with cold water before storing and used within a day or two. Whites should be covered and used within 2 days.

Sometimes a small blood or meat spot is found in eggs. Although these eggs rarely reach consumers, the spot can easily be removed. These eggs are perfectly wholesome and are just as good as any other eggs. These spots are not a sign the eggs are fertile (20).

The white, twisted rope-like strands of material found in raw eggs bother a lot of us. Known as "chalazas," these strands are highly concentrated and wholesome parts of the white and serve to anchor the yolk firmly in place. They are found in all eggs.

A cloudy or slightly milky white does not affect the quality of eggs we have stored for a short time. It's actually a perfectly normal characteristic and merely indicates that carbon dioxide, which is normally present in fresh eggs, has not yet escaped through the shell. The white becomes clearer the longer we store eggs.

Dried eggs should be kept in the refrigerator after a package has been opened. Unused portions should be kept in airtight containers. If we store them properly, dried eggs should keep their flavor and quality up to a year. If we want to keep dried eggs longer than one month, we should freeze them.

MILK

Few drinks are as refreshing as a glass of cold, fresh milk. Fluid milk, cream, and cultured milk products are at their best in flavor

STOCKING THE LARDER

and nutritive value when they are kept clean, cold, and tightly covered. Most milk products can be kept in the refrigerator for up to a week if they are kept tightly covered. However, for best flavor, we should use fluid milk and cream within 3 to 5 days and cultured products within 2 to 3 days.

Before we place a carton of milk in the refrigerator we should first rinse off the container with clean, cold water and dry with a clean cloth. Milk should be kept in the coldest part of the refrigerator.

If we keep all dairy products tightly covered or in sealed containers we will keep them from absorbing other odors in the refrigerator.

New milk should never be mixed with old unless we are going to use it immediately. Only the amount needed should be taken from the container. Once milk has been removed from the container it should never be returned. It should be stored in a separate container.

Milk and cream should never be exposed to the light for any length of time, because light destroys riboflavin and may even cause an off flavor. Milk and milk products should be refrigerated as soon as possible after we purchase them, and we should take them out of the refrigerator only long enough to pour out what we need.

Canned evaporated milk should be kept at room temperature until opened, and then should be treated as fresh, fluid milk.

Dry milk should be kept in its original container at temperatures below 75° F. or lower until reconstituted. Then the milk should be treated like fresh, fluid milk. The package of dry milk should be closed immediately after using. Unopened packages of nonfat dry milk can be kept for several months without being refrigerated. However, some flavor changes may occur if temperatures reach 90° F. or above for long periods of time. If milk powder is exposed to moisture in the air during storage, it may become lumpy and stale.

Fresh milk which has been frozen for a month or less can be used, although the flavor and appearance may be changed.

BUTTER AND MARGARINE

Butter and margarine should be stored in their original wrappers until we are ready to use them. They should be kept in the coldest part of the refrigerator. If butter and margarine are exposed to room temperatures for long periods of time, heat and light may turn them rancid. Butter and margarine should always be kept covered so they will not absorb flavors from other foods. If we do not plan to use butter and margarine within a short time, they should be frozen.

ICE CREAM

Frozen desserts and ice cream should be stored in tightly closed cartons at 0° F. or lower. If we must keep ice cream in the ice-cube compartment of a conventional one-door refrigerator, we should use it within one week. If we have to store ice cream in one of these compartments, we should cover opened cartons with protective wraps to hold moisture in.

Ice cream and ice milk which have thawed should never be refrozen.

CHEESE

The main problem we face when we store cheese is loss of moisture. Larger cheeses are much easier to keep, because most of them have a protective wax coating. We need only protect the cut side. Aluminum foil or plastic wrap are ideal for wrapping cheese, because they will stick to the sides of the cheese. We can also dip the cut side of cheese into melted paraffin.

All cheeses should be kept refrigerated. The length of time we can store cheese depends on the kind of cheese and the type of wrapping used. (The exception is processed cheese, which does not need to be refrigerated until after opening.)

Cottage and Ricotta should be used within a few days. Other soft varieties like cream and Neufchâtel should be used within two weeks. Hard cheese will keep up to several months, if it is protected from drying out and from mold contamination.

Almost all cheeses should be kept in their original container or wrappings if possible. However, blue-veined cheeses keep best in a domed container or if covered with a dampened cloth. These cheeses need some air to enable them to continue ripening. The domed container will tend to trap air, and the damp cloth will let air penetrate.

A good way to keep large pieces of cheese from molding is to brush them with vinegar before storing in a vinegar-soaked cloth.

Freezing is not recommended for most natural cheeses, since the texture may become crumbly and mealy. Soft cheeses tend to separate after they have thawed. Certain varieties can be frozen if the pieces are less than a pound in weight and an inch in thickness and if the pieces are carefully protected in moisture-vapor-proof materials. Varieties which can be frozen are Cheddar, Swiss, Edam, Gouda, Brick, Muenster, Port du Salut, Provolone, Mozzarella, and Camembert.

These cheeses should not be frozen for more than 6 months, and should be thawed in the refrigerator and used as soon as possible.

FRESH FRUIT

Those fresh fruits we have come to take for granted are precious commodities and extremely perishable.

Ripe, fresh fruit should be used as soon as possible. We should sort through any prepackaged fruit we have bought and discard any damaged pieces. Fruit can be washed, but it must be dried; dry fruit keeps longer.

Fruit needs tender loving care. Crushed or bruised tissues permit the entrance of spoilage organisms which quickly break down quality.

The best temperatures for ripening most fruits are between 60° and 70° F. They should never be ripened in direct sunlight. As soon as fruit is ripe, we should store it according to the directions for that individual type of fruit unless we intend to use it immediately.

Temperature and humidity are critical factors when we store fresh fruits. If fruits are kept in a plastic bag in the refrigerator, we should make a few small holes in the bag to let air in and moisture out.

Apples

Store fresh apples in the refrigerator to retain crispness. Ripe apples will usually keep a week or two. Hard, firm apples will keep several weeks in a cool, dark spot. Smaller apples keep longer than large ones. Small apples should be set back for later use.

Apricots

Store uncovered in the refrigerator and use within 3 to 5 days.

Bananas

Bananas will ripen quickly at room temperatures. The best temperature is somewhere between 60° to 70° F. If it is necessary to hold ripe bananas, we should cover them and store in the warmest part of the refrigerator and use them within a day or two.

Berries (Strawberries, Raspberries, Blueberries, Gooseberries, etc.)

They should be stored (unwashed and uncovered) in the refrigerator and used within 1 to 2 days.

Cranberries

Store unwashed in the refrigerator and use within one week.

Cherries (sweet)

Store unwashed and uncovered in the refrigerator and use within a day or two; however, they will keep for several days.

Citrus Fruits

Keep them in a cold room or in the refrigerator, uncovered. They will keep well for several weeks.

Dates

Store in a closed container in the refrigerator. They will last for many months if kept cold and fairly dry.

Figs

They will keep for only a day or two in the coldest part of the refrigerator.

Grapes

Store grapes unwashed and uncovered in the refrigerator and use within 3 to 5 days.

Melons

Keep in the refrigerator up to a week. Cut melons should be

STOCKING THE LARDER

covered with plastic or foil. Melons with a noticeable odor should be kept in a tightly closed plastic bag. Use cut melons in 3 to 5 days. Whole melons keep longer.

Nectarines

Store uncovered in the refrigerator. Use within 3 to 5 days.

Peaches

Store ripe peaches uncovered in the refrigerator and use within 3 to 5 days.

Pears

Ripen in a plastic bag on top of the refrigerator. Punch small holes in the bag. Store ripe pears, uncovered, in the refrigerator.

Pineapples

Allow to ripen at room temperature. Refrigerate ripe pineapples and use within 1 to 2 days.

Plums and Prunes

Store uncovered in the refrigerator and use within 3 to 5 days.

FRESH VEGETABLES

Vegetables must be stored in darkness and under cool, moist conditions with adequate ventilation. Most fresh vegetables should be kept in the lowest part of the refrigerator. (See Table 8.) Vegetables should be sorted before we store them, and we should discard any which show decay. They should be washed if they are very dirty, but should be dried well. Actually, the best practice is to wipe any dirt off with rough paper toweling before storing.

Artichokes

Refrigerate in a closed container or plastic bag for no more than 4 days.

Avocados

Ripen hard or firm avocados in a paper sack on top of the refrigerator. Store ripe avocados in the refrigerator uncovered and use within 3 to 5 days.

Asparagus

Spears should be wrapped in damp paper towels before placing in the crisper. Do not keep more than 1 day before using.

Beans

Store in a plastic bag in the refrigerator. Beans will keep for several days this way.

KEEPING FOOD SAFE

Brussels Sprouts
Store, uncovered and unwashed, in the refrigerator and use within 2 to 3 days.

Cabbage
Rinse with cool water, cover with plastic wrap, and store in the crisper of the refrigerator.

Carrots
They will keep well in the refrigerator but, for best nutrition and flavor, they should be used within a few days. Before storing, the tops should be cut off, then the carrots should be rinsed well and placed in a plastic bag.

Cauliflower
Refrigerate in a plastic bag and use within a few days.

Celery
Keep cold and moist. Slightly wilted celery can be freshened by placing the freshly trimmed butt end in water. Badly wilted celery cannot be revived.

Corn
Keep the green husks around the ears; keep refrigerated and use within 1 to 2 days.

Cucumbers
Keep in crisper of refrigerator for up to a week. Cut pieces should be wrapped in plastic film.

Eggplant
Refrigerate and use as soon as possible. It loses moisture rapidly.

Garlic
Store garlic in a capped jar in a cool, dry cupboard.

Greens (Mustard, Turnip, Chard, Collard, Dandelion, and Kale)
Store dry and unwashed in the crisper. Use within 2 to 3 days.

Lettuce
The rule of thumb: the crisper lettuce is, the longer it will keep. We should never store lettuce next to fruits like pears, plums, or apples or vegetables like avocados and tomatoes. They all expire as they ripen and give off ethylene gas. This makes lettuce develop rust spots.

STOCKING THE LARDER

Mushrooms

Store under slightly damp paper towels on a flat surface so air can circulate freely around them. Use within a few days.

Okra

Store in a plastic bag in the refrigerator crisper. Use within 1 or 2 days.

Onions

Store green onions in a plastic bag in the refrigerator. Dry onions should be stored in a cool, dark, dry place. Since dry onions and potatoes require the same type of storage conditions, many of us store them together. This is a mistake, because potatoes give off just enough moisture to make the onions sprout.

Peas

Keep them, unshelled, in the refrigerator and use within 1 to 2 days.

Potatoes

Store in a cool, dark, and dry place. Never store them in the refrigerator, however. At room temperatures, potatoes will keep up to 3 weeks. They will keep longer in a cool, dark basement. (See instructions for storing onions.)

Radishes

Remove the tops before refrigerating. They should be stored in a plastic bag and used within 3 to 5 days.

Sweet Potatoes

Store them in a dark, fairly humid place and they will keep up to 3 weeks.

Spinach

Store in a plastic bag and use within 1 to 2 days.

Squash

Summer or soft-rind squash are very perishable and should be kept covered in the refrigerator and used within 3 to 5 days. Winter or hard-rind squash will keep well in dark, cool, and dry areas such as a basement for up to 2 weeks.

Tomatoes

Store unwashed and uncovered in the refrigerator and use within 2 or 3 days.

Turnips and Rutabagas

Will keep well in a cool, humid place for 2 weeks.

FLOUR, CORNMEAL, MIXES, ETC.

More Americans than ever are turning to grain for nourishment. The flour and meal we've taken for granted for so many decades have achieved a new status, as well as price. These staples are the framework for many foods we bake, and since they contain significant amounts of protein and other nutrients, we need to make every effort to store them correctly.

They should be stored in a cool, dry place. Insects are attracted by the nourishment in flour and meal, so these grain products should always be stored in airtight containers. If flour is allowed to lose some of its natural moisture to the air, it will act like a sponge, and will absorb more than the normal amount of liquid from batter or dough.

For long-term storage, freezing is the best solution. Freezing is especially helpful for dark flours which contain oil-rich wheat germ. Wheat germ can become rancid if it is not stored properly. Flour and meals should be wrapped carefully in moisture-vapor-proof material before they are frozen.

Flour which has been exposed to high temperatures for lengthy periods may not be usable.

Containers should always be cleaned thoroughly before a new supply is added.

Mixes can be stored right on the kitchen shelves. However, ingredients in many mixes will deteriorate in time. Depending on what ingredients are included, shelf life may be as long as 1 year. In warm climates, shelf life will not be that long, however. If we think there may be insect infestation, mixes can be cold-treated in their original containers. (They can be exposed to 0° F. temperatures in a deep freeze for 3 to 4 days.)

SUGAR

All sugar should be stored in tightly covered containers at room temperatures. Ordinary brown sugar will often harden if air is allowed to penetrate to the sugar. Adding half an apple or a piece of fresh bread in the container provides moisture. However, we should check the bread or apple now and then for mold. Hardened

STOCKING THE LARDER

brown sugar can be softened by heating it in a slow oven; it will become harder than ever after it cools though, so it should be handled while it is still warm.

Maple sugar should be stored in a dry place at room temperature in a tightly covered container.

BAKING POWDER AND SALT

Keep covered and store at room temperature away from sources of heat.

FATS AND OILS

Most fats and oils need protection from air, heat, and light. Only a small amount should be kept at room temperature. The rest should be stored in the refrigerator to prevent any change in flavor. We should buy only in quantities we can use fairly soon. Some oils may cloud and solidify in the refrigerator, but they will clear up and become liquid again at room temperatures.

The first recognizable deterioration in fats such as lards, oils and shortenings is the development of rancidity. This is a change caused by oxidation. Vegetable oils, however, have a natural antioxidant. Since animal fats do not contain natural antioxidants to protect them, it is necessary for manufacturers to add chemical agents to delay the onset of rancidity.

Shortenings, either of vegetable origin or mixtures of animal and vegetable fats, keep well at room temperature. However, if a shortening is to be kept over a long period, we should store it in the refrigerator. Lard should be kept refrigerated. Shortening should be allowed to return to room temperature before we blend it into pastry ingredients.

Fats and oils in partially filled containers will keep longer if we transfer them to smaller containers in which there is little or no airspace.

Meat fat drippings should be stored tightly covered in the refrigerator and used within 2 weeks. They should never be allowed to stand for long periods at room temperature.

All homemade salad dressings must be refrigerated. We can safely keep purchased mayonnaise and ready-made salad dressings at room temperatures for several months before we open them. As soon as they are opened, however, they must be refrigerated.

All mayonnaise combinations and all foods mixed with them must also be kept refrigerated. They can become dangerously spoiled by microbial activity without showing any evidence of spoilage. It is safer to add vinegar or pickle to cooked foods that are to be mixed with mayonnaise or salad dressings, because the acid in either helps to keep down microbial growth. Even the addition of lemon juice helps. But adding the acid does not insure that the foods will be safe if they are not refrigerated.

SPICES

Spices sometimes may become infested with the so-called "store-grain" insects (those which attack flour and cornmeal). Members of the red pepper family and some of the dehydrated vegetables are especially apt to be infested. These pepper family spices are also likely to lose their good quality in 6 months. In warm, moist climates their deterioration is even more rapid.

We should always buy all spices in small quantities, because all of them will lose flavor if they are kept too long.

There isn't much point in adding spices to perk up recipes if the unique and characteristic flavors have deteriorated and vanished. We can assure maximum flavor and accent if we remember these points:

(1) Spices must be stored in tightly closed, screw-cap glass jars. Cans and cartons are less expensive, but they cannot be resealed properly.
(2) Our spices should be stored in a cool, dry location away from bright, artificial light or sunlight and away from sources of heat. We don't have to refrigerate spices, but unfortunately, the spot most often chosen for storage—right over the range—is the very worst choice.

BEVERAGES

Canned fruit and vegetable juices should be kept in cool, dark places. After opening, they may be covered and kept in the cans

STOCKING THE LARDER

and stored in the refrigerator. However, some flavor changes may occur, so most people prefer to transfer the juices to glass or plastic containers for storing. Once opened, juices should be kept at temperatures no higher than 40° F. (the normal refrigerator temperature).

Chocolate needs to be kept cool to prevent melting, and cocoa should be stored in a cool dry place; neither should be subjected to wide swings of temperatures. Tea should be covered tightly and stored at room temperatures.

Coffee is more difficult to keep. The oils and other constituents which provide coffee's tempting flavor and aroma are easily dissipated. If coffee is exposed to air for any length of time, these constituents will oxidize and the result will be vastly inferior taste and aroma.

Unfortunately, once a container of coffee has been opened, the loss speeds up. This loss can be slowed down, but not stopped, by keeping the tightly closed coffee container in the refrigerator. We should buy coffee in amounts small enough to use up quickly.

Wines require constant temperatures somewhere between 55° and 60° F. and should never be exposed to either daylight or vibration. (Of course, a bicycle ride to a picnic ground is another matter, because we'll be opening the wine soon.)

If we try to keep a few bottles of our favorite wines put back, we should pick a spot which meets the requirements of steady, cool temperatures and little or no vibration. (The wine cellar makes a lot of sense.)

Once a bottle of table wine has been opened, it will not keep well because of exposure to air. As a general rule, however, leftover white wines will store better than leftover red wines.

Leftover wines must be recorked as soon as possible, chilled, and used within a very few days. If we buy and open a half gallon of wine, we can keep the bouquet and taste more easily if we transfer the wine to smaller containers which can be tightly capped.

BREADS

Bread which will be used within 2 days should be kept in a bread box at room temperature.

If the weather is very warm and humid, bread should be kept in its original wrapper in the refrigerator to prevent mold. However, refrigeration hastens staling. Hard or crisp-crusted breads or rolls are best if eaten fresh. Whole wheat products are more difficult to store because of their high fat content. Breads may be frozen with satisfactory results. (See Chapter 4.)

All bread storage containers benefit from a weekly washing solution. (See Chapter 5.)

We don't have to throw out leftover bread. A glance at the prices of commercially prepared bread crumbs gives us motivation to dry and crush leftover bread to make our own crumbs. Leftover rolls, coffee cakes, and sweet rolls make excellent crumbs for bread puddings.

CEREALS

Few things are less palatable than stale or limp cereal. All cereals need to be covered to keep them from drying out or absorbing moisture.

To keep in the flavor of cereals after they have been opened, we should store them in cool, dry areas of our kitchens. If we open a cereal package carefully, we should be able to reseal it tightly.

Cereals should be stored in containers away from soaps or other products with strong odors.

HONEY, SYRUPS, JAMS AND JELLIES

Honeys and other syrups can be stored, unopened, at room temperature, but once they are opened they should be refrigerated unless we intend to use them quickly. If they form crystals in the cold air of our refrigerators, we can warm them gently, and they will return to their original consistency.

Honey stored over a long time may darken somewhat, but it will still be usable.

NUTS

Only nuts packed in vacuum-packed containers can remain fresh at room temperatures. Most nuts will mold or soon lose their flavor and even develop off flavor and darken, if they are not protected from air, heat, light, and excessive moisture. Rancidity is the worst enemy of nuts high in fat content. The fat reacts with oxygen from the air, causing a rancid odor and flavor. Nuts are also subject to mold. Moldy nuts should be discarded.

An excellent way to keep shelled, fresh nuts is in small containers in our freezers. Hard-frozen nuts can be chopped easily, and by the time we are ready to include them in a recipe, they will be thawed. Frozen nuts will keep their fresh qualities for up to 2 years if stored at 0° F. or lower.

Shelled nuts will retain top quality for up to 6 months if they are kept in tightly closed containers in our refrigerators.

One of the most delightful by-products of peanuts is peanut butter. Although unopened peanut butter will keep for a year at room temperatures, it should be refrigerated after it has been opened. We need to plan ahead a little and set peanut butter out to warm up a short time before we want to use it.

YEAST

If we buy yeast cakes, we should buy only the amount we intend to use. These little cakes are very perishable and must be refrigerated. They will keep well at temperatures of 40° F. for only a few days. When cake yeast is fresh, it is white, easily crumbled, and has a delightful aroma we all recognize. When it is stale, cake yeast dries out and becomes slimy, brown, and develops an unpleasant odor. Fresh cakes of yeast can be frozen, but must be used immediately after thawing. Never defrost and refreeze.

Active dry yeast will keep very well at room temperatures. Packages are dated with the month and year which indicate how long they can be used with satisfactory results. Active dry yeast can be kept for longer periods in the refrigerator or freezer. If storing large amounts, it is wise to keep a small portion on the shelf, stored at room temperatures for immediate use.

DRIED BEANS, PEAS, LEGUMES

These protein-packed foods should be kept in tightly covered containers in a cool, dry place (between temperatures of 50° to 70° F.). If the packages are not opened, they will maintain their quality several months.

After opening a package of beans, we should never mix the

STOCKING THE LARDER

contents with other packages of beans bought at separate times. Mixing beans which differ in age causes uneven cooking. (Older beans take longer to cook than fresher ones.) (32)

Once we have opened a package of dried beans, peas, or lentils, we should store the unused portion in a glass or metal jar with a tight-fitting lid.

DRIED FOODS

Most dried foods should be stored in a dark, cool place. (The refrigerator is ideal.) Tightly sealed containers are a must.

Dried foods can be compressed to take up less space for storage. They can be softened in the oven for half an hour, then packed into storage containers.

Once a package has been opened, the contents must be refrigerated.

CANNED PRODUCE

The ideal storage area for canned goods is one which is cool and dry. If canned goods are exposed to dampness, the cans may rust and the labels could come off or at least be stained. The lower the temperature, the slower will be the chemical changes which normally take place in canned foods during storage. These changes will eventually result in off colors and off flavors with accompanying loss of nutritive values and change of texture.

Storage temperatures above 100° F. can cause germination of thermophilic bacteria spores which will cause spoilage. Flat-sour and hard-swell spoilage may occur as a result. (A soil bacteria causes these types of spoilage. The foods usually have a sour, acid taste and an offensive odor. Some white sediment may also form in the bottom of jars.)

Most canned fruits and vegetables will keep up to a year, but all canned goods should be rotated as we use them.

Although cool storage is best for canned goods, they should not be allowed to freeze, because freezing causes undesirable changes in color and flavor. Canned cream-style corn, squash, cream soups, corned beef hash, and other foods containing a lot of starch show definite physical changes after freezing in cans.

STOCKING THE LARDER

If canned goods are exposed to a hard freeze, the cans may become distorted and sometimes the seams may even burst. Distorted cans may lead to leakage and then spoilage. When canned goods have frozen, we should check the containers very carefully after the foods have thawed. The ends of the cans should return to their normal flat appearance. If they do not, this means air has gotten into the cans, and the contents will spoil.

Accidental freezing of food will occur more rapidly in glass jars than in metal cans.

Damage to the outside of a can does not necessarily mean the contents are unsuitable for use. Rust or dents do not affect the quality of the food so long as the cans do not leak.

However, anytime there is a dent or rusted rims or seams, we must examine the containers carefully to make absolutely certain there is no puncture and leakage of air into the can. If a can is leaking, if the ends are bulged, or if the contents have an unnatural odor or appearance, the food should be thrown out and not tasted. Suspicious-looking foods should be destroyed so not even an animal can consume them.

We must never forget it is possible for canned vegetables to be contaminated with botulism without showing any signs of spoilage. This is particularly true of home-canned, low-acid vegetables and all meats. These home-canned foods must always be boiled for 20 minutes in an open kettle before we taste them. Boiling will destroy the deadly toxin produced by the organisms.

Occasionally some extremely acid foods, such as sauerkraut, will tend to react with the lining of cans if they are stored for lengthy periods. This can produce swelling. Since acid foods generally do not support the growth of spores which produce toxin, it is unlikely foods of this sort would cause botulism. However, the only way to tell if the swelling is caused by food spoilage organisms or by the reaction of the food to the can lining is by laboratory analysis. We should throw out all suspicious-looking foods.

We are becoming increasingly aware of how precious food is. But we must not forget that contamination, spoilage, and food illness may occur if we are not alert to the principles of storing food safely.

Table 7

FREEZER STORAGE

Product	Time
Breads, baked	2–4 months
Breads (yeast), unbaked	1 week
Cakes, cupcakes, baked	3–4 months
Cakes, unbaked	2 months
Cookies, baked or unbaked	4–6 months
Commercially frozen vegetables	
Asparagus	8–12 months
Snap beans	8–12 months
Lima beans	14–16 months
Corn	8–10 months
Corn (cut)	24 months
Spinach	14–16 months
Ice cream or ice milk	2 months
Pies	
Baked or unbaked	2–4 months
Dough only	4–6 months
Chiffon	2 months
Uncooked meat	
Beef	6–12 months
Pork	3–6 months
Lamb	6–9 months
Veal	6–9 months
Organs	3–4 months
Ground beef, veal, lamb	3–4 months
Ground pork	1–3 months
Smoked ham	2 months
Bacon	1 month
Combination dishes	2–3 months
Poultry (uncooked whole or cut up)	
Chicken	12 months
Turkey	12 months
Duck	6 months

STOCKING THE LARDER

Goose	6 months
Giblets	3 months
Seafood	
Fatty fish	6–8 months
Lean fish	10–12 months
Wild game animals	9 months
Wild game birds	9 months

TABLE 8

REFRIGERATOR STORAGE CHART

Freezing Compartment	Door
Fresh meat, poultry, fish—loosely wrapped Milk, cream, soft cheese—well covered	Eggs—only if covered Butter—only enough for immediate use
Butter, Margarine—tightly covered Eggs—in covered carton or other container Hard cheese—tightly covered	Hard cheese—tightly wrapped Small open jars—covered
Leftover cooked foods: meats, vegetables, fruits—covered Opened cans and bottles—covered Refrigerated rolls—unopened	Tall bottles, soft drinks, fruit juice—covered or corked
Ripe fresh fruits Ripe tomatoes, unhusked corn, lima beans, and peas in pods—uncovered	
Crisper or Hydrator Most fresh vegetables	

From: *Consumer News.* Aug. 15, 1973.

Table 9

TIME NEEDED TO THAW MEAT IN THE REFRIGERATOR

Cut	Thawing Time
BEEF	
Rolled rib roast	8–10 hrs. per lb.
Standing rib roast	7–8 hrs. per lb.
Rump	7–8 hrs. per lb.
Sirloin steak (1 inch)	8–10 hrs. per lb.
Round steak	5–6 hrs. (total thawing time)
Ground beef (1-lb. pack)	18–20 hrs. (total thawing time)
Patties (separated)	5–6 hrs. (total thawing time)
Stew meat (1-lb. pack)	18–20 hrs. (total thawing time)
PORK	
Chops (¾-inch thick)	4–5 hrs. (total thawing time)
Cutlets	7–8 hrs. per lb.
Steak	7–8 hrs. per lb.
Sausage	18–30 hrs. (total thawing time)
VEAL or LAMB	
Chops	7–8 hrs. (total thawing time)
Leg	7–8 hrs. (total thawing time)

TIME NEEDED TO THAW POULTRY IN THE REFRIGERATOR

Weight and Type	Thawing Time
3½-lb. (cut up)	10–20 hours (in unopened package)
5 lbs. (whole)	1–2 days (in unopened package)
	4–5 hours (under running water)
8–11 lbs.	1–2 days (in unopened package)
11–18 lbs.	2–3 days (in unopened package)
18–24 lbs.	3–4 days (in unopened package)
Giblets	Overnight (in unopened package)

From: *The Time of Your Life.* Sears Roebuck & Company 586525. Revision B.

9

Investigation and Control of Food

Vichyssoise is one of those sophisticated, elegant cold soups which gourmets favor. On the evening of June 29, 1971, in Bedford, New York, Samuel Cochran, Jr., and his wife sat down to dine on canned vichyssoise. The soup had been manufactured and distributed by a small, but reputable firm, Bon Vivant.

After a night of terror and panic, the banker was dead. His wife was in grave condition in a local hospital. There had been botulinum toxin in the can of vichyssoise (6).

And before many more months had passed, a giant food company had recalled millions of cans of suspected soup.

Next, more than 65 brands of canned mushrooms were involved in 12 recalls by the Food and Drug Administration. Once again, the suspect was the deadly botulinum toxin (21).

For years, authorities had assured the American public that botulinum toxin was primarily a danger in home-canned foods. Suddenly, here was concrete evidence that the commercial canning industry had serious botulinum problems of its own.

In September 1972, one hundred and twenty-eight persons became ill after eating food prepared in an Owatonna, Minnesota, bar. All the victims had either eaten food catered by the bar at a sales picnic or had eaten at the bar itself. In fact, some of those who attended the picnic had enjoyed the food so much, they went to the bar the next day and ordered more.

The malady was diagnosed by state health investigators as salmonellosis (10).

So the trusting American consumer is no longer as confident that the food bought in stores or restaurants is pure and wholesome. Millions of us have demanded that our governmental agencies crack down on the food industry.

We want to know just who is making certain that the ingredients used in foods are really pure and wholesome. And we wonder who makes sure only safe food is produced in the big, corporate kitchens which provide millions of us with convenience food.

Traditionally, public indignation and concern have been the most powerful forces in producing constructive change by governmental agencies.

There were no truly effective controls on food safety until the mid-1800s, when preventive measures designed to stamp out diseases among livestock were adopted.

Then in the early 1900s an American author named Upton Sinclair began moseying around the Chicago packing-house district. His controversial novel *The Jungle* was written about the conditions he found there.

The result was a public clamor so great that the United States Congress felt compelled to pass the Meat Inspection Act of 1906 (20).

And ever since President Theodore Roosevelt signed both the Pure Food and Drug Act and the Meat Inspection Act, the Federal Government has been checking on the safety and wholesomeness of our food.

How ironic, then, that in a country as progressive as ours, and in an era in which scientific knowledge abounds, food inspection practices should still be a problem of grave concern.

But to its credit, the entire food industry is making every effort to see that their goods are not contaminated with dangerous organisms or toxins.

According to Food and Drug Administration officials, our food is safer now than it has ever been before (25). Yet word of new

INVESTIGATION AND CONTROL OF FOOD

food recalls, and horror stories about filth in foods and food plants continues to keep the public concerned about commercially prepared food.

For instance, in the early part of the 1970s, the American consumer learned that the FDA "filth allowances" permit one rodent pellet per pint of wheat and 10 fly eggs per 8½-ounce can of fruit juice. Peanut butter is acceptable if it contains no more than 50 insect fragments or 2 rodent hairs per 3⅓ ounces.

FDA's Dr. Virgil O. Wodicka, Director of the Bureau of Foods, explained that the issue of unavoidable defects in food for human use is difficult to discuss.

The rodent hairs and the insect parts may bother us esthetically, yet they present no health hazards, the FDA firmly maintains.

However, this book is not intended as an exposé of laxity in either the government or industry.

The sheer numbers of companies involved means that no one agency or even groups of agencies can possibly provide continuous inspection at all food plants.

Fortunately, however, professional groups such as the National Canners Association have traditionally endeavored to promote and enforce high standards within the industry. (This group has recently stepped up its efforts toward the adoption of stricter rules and closer cooperation with federal agencies in inspection and control.)

As the public has become more alarmed about the cleanliness and safety of commercially prepared food, government agencies on all levels have increased inspection activities. Beginning in 1972, the Department of Health, Education and Welfare's Food and Drug Administration implemented a new form of surveillance. One result was 12 recalls of canned mushrooms (2).

Laws Which Protect Our Foods

FOOD, DRUG AND COSMETIC ACT

The Federal Food, Drug and Cosmetic Act, enacted in 1938, replaced the first food and drug act which President Roosevelt signed into law in 1906. Since 1938 there have been many amendments to the original act, and all have been designed to assure greater safety for our food. The basic intent of the law as it pertains to food is to assure us that:

(1) Our foods are pure and wholesome.
(2) Our foods are safe to eat.
(3) They are produced under sanitary conditions.
(4) The packaging and labeling of these foods are truthful and informative.

It deals in part with the prohibition in interstate commerce of adulterated and misbranded food and provides authority for review of food and color additives.

Other amendments in the future will undoubtedly strengthen and broaden the law, including requirements that food processors maintain more complete records concerning possible adulteration or misbranding and mandatory registration with the FDA. Other amendments may also require standardized labeling of ingredients and grant the power of administrative detention of foods by the FDA.

This law is enforced by several parts of the FDA, including the Bureau of Foods. The FDA has 19 district offices in cities across the United States which provide enforcement of all laws on a local level (24).

THE FAIR PACKAGING AND LABELING ACT

Enforced by the Federal Trade Commission and the FDA, this law requires that labels inform the consumer of the true ingredients and weight of the package. FDA's role is limited to foods, drugs, and cosmetics.

THE PUBLIC HEALTH SERVICE ACT

The FDA's Bureau of Foods operates under this law to assure the safety of pasteurized milk and shellfish.

THE TEA IMPORTATION ACT

This law assures us that all teas imported into the United States meet the quality standards set by the U. S. Board of Tea Experts.

INVESTIGATION AND CONTROL OF FOOD

Samples of imported teas are examined before being allowed on the market.

THE WHOLESOME MEAT ACT AND THE WHOLESOME POULTRY PRODUCTS INSPECTION ACT

Both laws are enforced by the USDA's Animal and Plant Health Inspection Service. Federal inspection is required for all meat and poultry prepared in plants that sell their products in interstate and foreign commerce. It includes inspection of cattle, calves, sheep, hogs, goats, horses, chickens, turkeys, ducks, geese, and guinea fowl.

Farmers or poultry producers who ship meat or poultry products which are unwholesome or "otherwise unfit" for human consumption across state lines are subject to a fine and/or imprisonment.

Meat and poultry plants which sell all of their products within the state in which they are produced are not subject to federal inspection. However, state inspection laws must be at least as strict as federal laws. In some cases, they are even more strict.

THE IMPORT MEAT ACT

Under this law, all meat and poultry exported from foreign countries to the United States must meet the same requirements of sanitation and inspection that are applied to meat and poultry produced in the United States. A foreign country must apply for eligibility to export meat and poultry to the United States. Then the USDA makes sure that country's inspection laws, regulations, procedures, personnel, and operations meet U.S. standards.

IMPORT MILK ACT

This law regulates the importation of milk and cream into the United States in order to promote our own domestic dairy industry. It also is designed to protect public health and to help assure us of a milk supply that is clean, fresh, and wholesome.

EGG PRODUCT INSPECTION ACT

All egg products (frozen and dried) must be processed in an official plant under USDA inspection. USDA inspectors also must check all producers who candle or sell eggs. In addition, no dirty or "checked" eggs can be sold to bakeries, restaurants, or institutions. The act specified these firms or agencies must buy Grade B eggs (or better).

WHO INSPECTS WHAT?

The names of agencies change from time to time. We are discussing the functions of various agencies. (The titles may not remain the same.)

FEDERAL AGENCIES

The agencies of the Federal Government which are directly involved in protecting our food supply are the Food and Drug Administration of the Department of Health, Education and Welfare, and the Meat and Poultry Inspection Program and the Agricultural Marketing Service, both segments of the vast and complex United States Department of Agriculture. The United States Public Health Service and the Armed Services are also involved in food inspection and protection.

These agencies must confine their enforcement and regulation to:

(1) Food shipped across state lines.
(2) Food which is produced in territories of the United States.
(3) Food shipped into territories of the United States.

Federal inspection by the USDA is required of all plants which ship meat or poultry across state lines or in foreign commerce. The Wholesome Meat Act of 1967 and the Wholesome Poultry Products Act of 1968 brought many small plants under inspection.

The Food and Drug Administration enforces, among other laws, the Federal Food, Drug and Cosmetic Act, the Tea Act, the Import Milk Act, and the Fair Packaging and Labeling Act.

The FDA enforces part of the Public Health Service Act when it assures the safety of pasteurized milk and shellfish and the sanitation of food services, food, water, and sanitary facilities for travelers on trains, planes, and buses.

INVESTIGATION AND CONTROL OF FOOD

In addition, the FDA also provides advice and technical data to the food industry, interested law enforcement agencies, and the public. The agency also conducts research in the field of food safety, makes regulatory checks, and sets specific standards for compliance.

The Animal and Plant Health Inspection Service enforces the Meat Inspection Act and the Imported Meat Act. The Agricultural Marketing Service provides inspection and grading of fresh produce, poultry, and dairy products.

The United States Health Service is not an enforcement agency in the strict sense of the word, but makes recommendations and aids other agencies. The Public Health Service is also involved in giving publicity to federal enforcement and regulations concerning the safety of food.

The Armed Services have their own standards and laboratories. During World War II, when the Armed Services issued specific instructions for poultry inspection, both the public and the poultry industry became familiar with inspection procedures. By the mid-1950s there was enough public demand to motivate both the government and industry to provide poultry inspection.

New pronouncements and developments in the field of food inspection, regulation, and enforcement are published in the Federal Register. Periodically, the Agricultural Marketing Service publishes papers which also highlight recent recommendations concerning inspection and grading of foods.

However, most of us are unaware of these developments. The information which flows from these various governmental agencies is available to the public, but few of us order each new government publication or bulletin dealing with the general subject of food safety. And we should not have to watch the food industry closely. Our governmental agencies such as the FDA and the USDA's Meat and Poultry Inspection Program and Agricultural Marketing Service were created to do this for us. We should be able to assume that food processed in plants under their jurisdiction is indeed pure and wholesome.

STATE AGENCIES

State food laws are generally enforced through the state departments of public health, agriculture, or sanitary engineering.

The underlying relationship between state and federal laws provides that state laws be at least as strict as federal laws. Often state and local laws are more strict than their federal counterparts.

MUNICIPAL AGENCIES

City or town food laws are usually enforced by the local boards of health. States and cities have food laws which are usually modeled on federal laws or recommendations.

There is often an overlapping of concern and jurisdiction, which should provide us with even more assurance of safe food.

A good example of how municipal and state inspection agencies work with federal agencies is in the inspection of milk.

The Food and Drug Administration is charged with formulating definitions for dairy products and is involved in the supervision of all interstate milk shipments.

Then at state level, milk control is generally a function of the state department of health or the state department of agriculture. These agencies cooperate with local governments to initiate and enforce the milk control programs. The state agencies also act as the main liaison between local governments when problems arise on interstate milk shipments.

And whenever a milk supply is not governed by a local ordinance, the state agency can assume control (20).

As a rule, municipalities or counties have the last say in the control of our milk supplies. And there is a growing tendency for cities, towns, or suburbs to share the control of milk supplies. Thus, inspection of farms and enforcement of regulations may be conducted by the local agency nearest the dairy farm, and the more distant government agency accepts the results of the jurisdiction of the other.

This has involved close cooperation between governments in developing uniform standards and ordinances. Adoption of the Public Health Service's Milk Ordinance makes uniformity possible (20).

COMMERCIAL AGENCIES

Trade associations or institutes may make recommendations or even regulate within their own industries. For example, the National Canners Association has set bacteriological standards for sugar and starch for canning; the American Dry Milk Institute has established bacterial standards for dry milk; the American Bottlers of Carbonated Beverages has bacterial standards for sweetening agents used in soft drinks, etc.

PROFESSIONAL SOCIETIES

The American Public Health Association has published many of the recommended and official methods for the bacteriological examination of foods. The International Association of Milk and Food Sanitarians has published recommended methods for the investigation of outbreaks of food illnesses, etc.

PRIVATE AGENCIES

Several private agencies approve and list tested foods. One example is the consumer-oriented Good Housekeeping Institute.

Inspections and Controls

When we hear stories such as the one about the contaminated vichyssoise, we cannot help but wonder if there are killers lurking in the cans of food on our own cupboard shelves.

While there is no doubt in our minds about the good intent of federal agencies, a Food and Drug Administration official did admit that the Bon Vivant plant had not come under FDA scrutiny since 1967, almost four years before the tragedy in the Samuel Cochran home (4).

FDA officials admitted they didn't know the plant existed, even though the firm had more than a $10 million annual sales volume (4).

Dr. Howard Bauman, of the Grocery Manufacturers of America, told a Congressional hearing that the food industry is probably the only one where someone could enter a business and start selling products without much in the way of any control, other than occasional inspection by Food and Drug (9).

The problem of adulterated food in the United States is significant. In 1973, for example, the Food and Drug Administration recorded 3,020 regulatory actions which included recalls, seizures, and import detentions against adulterated food (9).

Food and Drug Administration officials have complained for years that the agency had neither the manpower nor the money to do a more thorough job.

In the past, once Food and Drug Administration inspectors gave a food plant a clean bill of health, it was often years before subsequent inspections were conducted. However, all complaints were acted upon, and FDA checked regularly on plants which had received bad reports in previous years.

Because more investigators have been added to Food and Drug Administration inspection rolls, the agency is moving in the direction it previously took in the regulation and inspection of drug manufacturers.

Registration with the FDA by food manufacturers seems to be the answer, according to agency officials. In fact, mandatory registration by food manufacturers and processors can be the means to developing better sanitary control, fewer violations, and ultimately cleaner food.

Approximately six months before the first can of contaminated mushrooms was discovered by the Food and Drug Administration, that agency had initiated intensive investigative activity centering on products manufactured by only a few canned food producers (2).

The mushroom processors were routinely included, because mushrooms are a low-acid food. (And we know now that these foods have a high contamination risk.)

Then in February 1973, contamination was found in mushrooms produced by the United Canning Company, located in East Palestine, Ohio. Furthermore, it was discovered that some of the mushrooms had been used in frozen products. So both the mushrooms and the frozen foods had to be recalled (2).

Fearing the worst, FDA officials then conducted an inspection of all canned mushrooms. In all, 42 firms were inspected, and serious "deficiencies" were found in half the industry.

The next step involved examination of all domestic and imported canned mushrooms which involved millions of cans in more than 9,000 warehouses. By the end of February 1974, one out of five U.S. mushroom canners had been involved in recalls (2).

More than $6 million went into the mushroom inspection program. And though the intensive program drained the FDA's financial resources, agency officials were confident they had at last made mushrooms safe and wholesome for American consumers.

The overall result of the mushroom crisis was to make federal agencies aware of the importance of identifying and correcting problems before they occur.

The FDA does not have the legal power to order a recall. All it can do is request one. However, the effectiveness of national publicity of a recall request usually is enough to initiate action. But the agency can seize adulterated food in interstate transactions through a federal court order.

However, in cases it deems serious enough, the FDA can also work with the Justice Department to bring charges or to apply for an injunction to halt a plant's production.

The law offers equal protection to the food industry, however. Peter Stevenson, Chief of the General Environmental Health Service, Department of Health, Denver, Colorado, has emphasized that a food plant operator has a right to know what charges the governmental agencies are filing. These operators should be given time to correct the condition or to appeal. Then if the condition is not corrected, or if the appeal is denied, the FDA has the right to take corrective action.

By the FDA's own calculation in the early 1970s, it was responsible for inspecting 32,000 food manufacturers and processors and 28,000 warehouses, grain elevators, repacking and relabeling plants.

These figures did not include all the food plants in the FDA's jurisdiction, because anyone could start up in the interstate food business without either being licensed by the FDA or registering with it.

The FDA made 9,000 plant inspections in fiscal 1972, and almost the same number the previous year. This means that the average plant was visited only once in 6.7 years.

In 1971, Dr. Charles C. Edwards, FDA commissioner, testified before a Congressional committee. He estimated the agency would need a sixfold increase in men and money to do its job adequately. (The FDA did get 300 more inspectors in 1973.)

Even though the FDA's manpower has been increased, the number of inspectors still does not begin to compare with the vast number of USDA inspectors.

However, the FDA and the Department of Agriculture use different inspection procedures. Traditionally, the FDA has used a spot-check system, sending an inspector around from time to time.

The USDA by law has to inspect every carcass and every bird to insure its wholesomeness, which has meant a system of continuous inspection.

Dramatic changes have occurred in meat and poultry inspection activities within the past twenty years. Until the last few decades, most inspectors relied almost entirely on their senses of sight, touch, and smell to detect abnormalities in meat or poultry. In other words, if the inspector didn't see, feel, or smell something abnormal, it didn't exist (26).

Today, the Department of Agriculture's meat and poultry inspection teams can detect a substance in animal tissues in terms of parts per billion or even parts per trillion (26).

In 1967 the Food and Drug Administration joined with the Department of Agriculture in a program to eliminate *Salmonella* contamination from animal by-products used in animal feeds. The USDA coordinated a voluntary program in more than 700 rendering plants. The FDA inspected the remaining 125 plants that did not participate in the voluntary program.

After five years of the program, out of a total of 741 cooperating plants, 196 had as few as two negative tests during 1972.

But the USDA and FDA decided to go even further in trying

to stamp out *Salmonella* and other organisms from the food chain.

A task force was set up in each agency to make independent studies of the needs and the available resources to fill those needs.

Both agencies have tightened up requirements for *Salmonella* control in processing plants.

The investigator representing the USDA, who inspects meat and poultry products, checks all animals and birds both before and after slaughter for abnormalities. He also checks internal organs for signs of disease and the eviscerated carcass as it progresses along the processing line.

Meat and poultry products are reinspected by USDA at processing plants to make sure they are still fresh and wholesome and that nothing has happened to contaminate them.

The Department of Agriculture has also built a highly skilled staff to back up its inspectors in the field.

One of the newer support functions in the scientific area deals with potentially hazardous residues being found more often in animal tissue.

A specialized staff has been developed whose function is to deal with the problem and its ramifications.

The USDA's Animal and Plant Health Inspection Service sets up regulations and offers guidelines for the facilities used by food processors. These guidelines have been developed during more than sixty years of inspection experience (16).

The guidelines are flexible, but they make sure that all plants produce sanitary, wholesome products under rules common to public health requirements for food processing.

For instance, in meat and poultry processing plants equipment must be easily dismantled—like the beaters that drop out of our mixers with the push of a button—so they can be cleaned easily.

All plant layouts, equipment, and facilities must be approved by USDA to make sure they meet all requirements before a plant is granted federal inspection.

An inspector examines the equipment in a plant to make sure that it is clean before work begins each day. He then checks all the operations in the slaughtering or processing plant to be certain products are handled properly and that equipment is cleaned as often as necessary.

To double-check the cleanliness and safety of products, the inspector can send samples to laboratories to have them tested by microbiologists, chemists, or pathologists (16).

In 1925 a typhoid fever outbreak, involving more than 1,500 cases and 150 deaths, was traced to consumption of contaminated oysters.

In response to requests from state and local health officials and from the oyster industry, the Public Health Service developed criteria for public health protection of these shellfish during growing, harvesting, and processing (20).

But even though the inspection procedures are well outlined, there is little the FDA or USDA can do to control the contamination of fish or seafood from polluted waters or to prevent the illegal sale of contaminated products.

In November 1973 more than 180 cases of hepatitis in the Houston, Texas, area were traced to eating contaminated oysters taken illegally from posted Louisiana waters.

The FDA inspects fish plants on a spot-check basis, and the Department of Interior provides a voluntary service which plant operators may buy.

There is also an ongoing state certification program of shellfish breeding waters and processing plants. But these programs do not keep poachers from selling contaminated food caught in posted waters.

However, on an overall basis, the picture is improving as far as the quality of our food is concerned.

Dr. Virgil O. Wodicka, Director of the FDA's Bureau of Foods, says the American food supply is safer now than it has ever been before, and as safe as we know how to make it. Agricultural sanitary practices are probably better now than ever before. Pesticides, about which there has been a great deal of concern, are probably safer than the ones used by our grandparents. The whole trend of the times has been to increase safety rather than to impair it (25).

INVESTIGATION AND CONTROL OF FOOD

Grading

If we had to choose for ourselves which fruit or vegetables were most desirable, or which turkey had the tenderest meat, we would be in a constant dither.

We are fortunate that the USDA grades foods for us. Grades are measures of quality. If a food has been scrutinized by a government grader, it may carry the official shield mark. However, we will not find the official grading shield on all foods, because federal laws do not require it.

Grading is a service which the USDA's Agricultural Marketing Service provides for those packers or processors who want to use it and who can meet the requirements for using it.

The foods we are most likely to find carrying the USDA grade shield are beef, lamb, chicken, turkey, butter, and eggs. A grade mark on one of these foods means it has been examined by an expert government grader, and he has certified that it measures up to a specific quality standard. Fresh fruits and vegetables are often graded but seldom bear the shield marks in retail outlets.

Into the Future

Though there is criticism of both governmental agencies and the food industry from time to time, the fact remains that the food we eat today is far safer than it has ever been in the past.

But that is not enough. As consumers we cannot relax our pressure and influence on our Senators and Representatives to provide the means for continually improving the food supply. And we must never accept inferior products from the food industry. We are partially responsible for inferior food if we are willing to buy it.

A significant and rather far-reaching hearing was held in the last quarter of 1973 by the Senate Committee on Labor and Public Welfare. The hearing centered on the need to protect our food supply by means of surveillance regulations for the detection and prevention of adulterated food. Considerable evidence was accumulated by the Committee which indicated a definite need for the registration of all food processors.

Undoubtedly, registration and more thorough surveillance of food processors will beef up inspection programs in the future. An encouraging aspect of the hearings was the testimony by industry groups such as the Associated Retail Bakers of America, the Shellfish Institute of North America, and others which indicate broad industry support of both registration and increased surveillance.

However, the drive for safer foods, and therefore fewer foodborne illnesses, is not being confined to industry and governmental agencies.

In August 1973 the U. S. Departments of Agriculture and Health, Education and Welfare recommended consumer education as the most effective way to combat foodborne illnesses.

Despite industrial precautions and controls when it comes to protecting food, the fact remains that improper handling of food at the retail level and in the home is a major source of food illness (26).

For example, *Salmonella* may be present in a chicken or turkey approved by an inspector. The Department of Agriculture's seal means that the meat is wholesome only if properly cooked and handled by the consumer.

There will undoubtedly be further expansion and coordination of an intensive consumer education campaign aimed at eliminating careless food handling practices in our homes and in food service establishments.

There is to be a continuation of a cooperative federal-state-industry program coordinated by the Food and Drug Administration to eliminate *Salmonella* from rendered animal by-products used in animal feeds.

USDA has pledged modification of processing procedures and facilities in meat and poultry plants under its inspection to reduce bacterial cross-contamination of products and equipment.

Also on tap is intensified support of industry and USDA financed research aimed at controlling and eliminating *Salmonella* throughout the food chain.

In the future there will also be development of model ordinances governing sanitation and food handling in retail stores, food service institutions, and the food transportation industry (19).

INVESTIGATION AND CONTROL OF FOOD

Investigation of Foodborne Disease Outbreaks

Foodborne diseases include those illnesses resulting from consumption of any solid food or of milk, water, or other beverages we may consume. (The more important diseases and their causes were explained in Chapter 1.)

From the public health viewpoint, the main purposes of an investigation of an outbreak of foodborne disease are to determine:

(1) How the foodstuff became contaminated.
(2) If growth of a toxigenic or infectious organism was involved.
(3) To find how such growth could take place.
(4) What measures may be taken to prevent a repetition of the same set of circumstances.

Prompt investigation is very important, so that the outbreak can be limited and so the physician may be assisted in treating the victims. Other factors which are important are:

(1) Exact location and identification of the causative agent.
(2) Establishment of the means of transmission.
(3) Demonstration of the opportunity for growth of the pathogenic organism.
(4) In instances of infections, proof that the pathogen has infected the victims is needed.

When we read about an outbreak of food illness, we often learn something about how it may have occurred, and therefore we are educated about how to avoid outbreaks in our own homes.

There was a time when we did not hear very much about outbreaks of foodborne illness. For decades, food was canned haphazardly, and there were many cases of serious foodborne illness.

But we didn't hear about these victims very often, because communication between cities, regions, and countries was not as swift and effective as it is today.

Today, when a serious outbreak of food illness occurs, an investigation team will interview everyone they can find who consumed the suspected foods. The team, which will vary with the public health department concerned, will talk with both ill and healthy people who ate the food. These teams will usually consist of a leader, and field and laboratory groups.

The field group will be responsible for the actual interviews, for providing treatment and nursing of the victims, and for personnel at the place of exposure to the disease. The field group will also collect samples of suspect foods and specimens from patients or food handlers when necessary. They will inspect the premises where the foods were stored, prepared, and served and will determine where the suspected foods were bought. Then they must investigate conditions at the retail or wholesale outlet and finally fill out reports on all their activities.

Meanwhile, the laboratory group makes necessary tests. The team leader interprets all the data.

If there are no samples of leftover foods and beverages, investigators may have to resort to rinsings, garbage, or food handled the same way and under the same circumstances that the suspected food was prepared.

The type of specimen investigators collect depends on the illness concerned. Cultures from the nose or throat or from sores on the skin of food handlers are used to determine how "Staph" is spread.

Fecal specimens are used for testing for *Salmonella* or *Shigella,* and vomitus is usually tested when investigators suspect chemical poisoning.

We have learned how to prevent outbreaks of some of the more important foodborne illnesses.

We know now that in order to keep foods safe, we must keep them clean, we must buy uncontaminated foods, we must keep food away from pests, and we must practice good hygienic habits.

We have learned we should avoid any conditions which might allow microorganisms to grow and multiply, and to reject any foods we suspect of contamination.

Since many cases of food illness have been traced to places where large groups of people have been fed (such as restaurants, catered dinners, and large picnics), food sanitation experts recommend we use a basic check list when we eat out.

The chances are good that the restaurants in our areas are safe if our communities maintain adequate inspection systems.

However, an American Medical Association symposium on the subject of foodborne illness revealed that in many areas of the United States these inspection codes were out of date or inadequate and that the inspections were infrequent or incomplete.

When we pick up a foodborne disease in a restaurant, we usually become ill only after we are back home. We probably will not connect the illness with the restaurant, particularly if we become sick after more than a few hours have passed.

Because we seldom are able to see a restaurant's kitchen, we must rely on other clues to determine if a restaurant is safe to eat in. Some of these clues are:

(1) Are the dishes cracked or chipped, and is the silverware pitted? (Pathogens can collect in cracks or pits.)
(2) Is the general appearance of the restaurant shoddy or dirty?
(3) Can any of the foods be touched or handled by customers? (Rolls, pies, and other foods should be kept out of the reach of customers in a serving line.)
(4) Do restaurant employees handle food and utensils in ways which can lead to contamination? (Do they use their hands when they could use utensils, or do they handle the food surfaces of plates, the tops of glasses, or the parts of silverware which touch customers' lips?)
(5) Are the employees neatly dressed? (Dirty or soiled uniforms and dirty hands are other potentially dangerous sources of contamination.)
(6) Are foods which are supposed to be hot served hot? Are foods which should be cold served cold? (If gravies are not hot, and if potato salad is not cold, there is reason for concern. Moderate temperatures are dangerous, because they encourage microbial growth.)

Anytime we suspect a serious food illness, we should:

(1) Contact our family physician.
(2) See that the food illness is reported to health authorities.
(3) Make samples of the food available to investigation teams.
(4) Take the necessary precautions to make sure it doesn't happen again.

How We Can Help the FDA

Anytime we come across a food that we suspect may be mislabeled, unsanitary, or harmful, we should report it to the Food and Drug Administration.

The information which we can supply to the FDA about a food of this sort can often lead to detection and correction of a serious violation of a food law. Many products have been recalled or removed from the market because of action initiated by a consumer.

Of course, the FDA cannot take action solely on the basis of our complaints, but at least the agency will investigate the situation. Then if a hazard is found, the FDA will try to remedy the situation (23).

Before we report any hazardous food to the FDA, however, we should ask ourselves these questions:

(1) Have we used the product as labeled?
(2) Did we follow the instructions carefully?
(3) Did an allergy contribute to the bad effect?
(4) Was the product old before we opened it?

Complaints can be made in writing or by telephone to the nearest FDA field office or resident inspection station. The FDA has 10 regional offices, 19 district offices, and 97 resident inspection stations throughout the United States.

The address and telephone number of the nearest FDA office will be listed under U. S. Government, Department of Health, Education and Welfare, Food and Drug Administration in all metropolitan telephone directories.

Or, complaints can be sent directly to FDA headquarters. The

INVESTIGATION AND CONTROL OF FOOD

address is Food and Drug Administration, 5600 Fishers Lane, Rockville, Maryland 20852. The complaint will reach the right person.

HOW TO REPORT

A complaint should be filed as soon as possible after an incident occurs. We should give our name, address, telephone number, and directions on how to get to our home or place of business, and state clearly what appears to be wrong.

We should describe in as much detail as possible the label on the product and give any code marks that appeared on the container. For example, markings on canned goods are usually embossed or stamped on the lid.

We should give the name and address of the store where we bought the food and the date we bought it.

We should save any remains of the food or the empty container for our doctor's guidance or for possible examination by the FDA. Furthermore, we should save any unopened containers of the product which we bought at the same time and place.

And, of course, if we are seriously ill, we should see our physician at once.

We should follow through by reporting the suspected product to the manufacturer, packer, distributor, and to the store where we bought it. The majority of businesses, manufacturers, or processors will appreciate our telling them about contaminated products which have been processed at their plants or which have been sold by their firms (23).

The FDA has limited jurisdiction over certain consumer products. If we have complaints about any of the following, these are the federal agencies to inform:

(1) Suspected false advertising—Federal Trade Commission, Washington, D.C.
(2) Meat and poultry products—U. S. Department of Agriculture, Washington, D.C.
(3) Fish and seafood—Mr. J. R. Brooker, Acting Chief, National Marine Fisheries Service, Division of Fisheries

Products Inspection, 1801 N. Moore Street, Arlington, Virginia 22209. Send copies of the letter to the Food Bureau, Food and Drug Administration, Washington, D.C.
(4) Sanitation of restaurants—local health authorities.
(5) Products made and sold exclusively within a state—local or state health department or similar law enforcement agency.
(6) Accidental poisonings—Poison Control Centers.
(7) Pesticides and water pollution—Environmental Protection Agency.

In the final analysis, food safety depends upon us all, industry, government, and consumer. With cooperative effort, we may be able to eliminate most food illnesses in our lifetimes.

Bibliography

CHAPTER 1

1. Anonymous. September 1971. "Botulism: An Unnecessary Menace." *Consumer Reports*, Vol. 36, No. 9.
2. Anonymous. July 19, 1971. "Death in Cans." *Time*.
3. Anonymous. August 1970. *Food Poisoning—Salmonellosis and Staphylococcus Poisoning.* Consumer Bulletin. Vol. 53, No. 8. U. S. Printing Office, Washington, D.C.
4. Center for Disease Control. June 30, 1973. "Epidemiologic Notes and Reports, Shigellosis on a Caribbean Cruise Ship." *Morbidity and Mortality*, Vol. 22, No. 26. Atlanta, Georgia.
5. Center for Disease Control. 1970. "Foodborne Outbreaks." *Annual Summary*. Atlanta, Georgia.
6. Department of Health, Education and Welfare. "Glazes and Decals on Dinnerware." No. x (FDA) 7301020. U. S. Government Printing Office. Washington, D.C.
7. Deutsch, Ronald M. 1971. *The Family Guide to Better Food and Better Health.* Creative Home Library. Meredith Corporation, Des Moines, Iowa.
8. Frazier, W. C. 1967. *Food Microbiology.* McGraw-Hill Book Company, New York.
9. Gifft, H. H.; M. B. Washbon, and G. G. Harrison. 1972. *Nutrition, Behavior and Change.* Prentice-Hall, Inc., Englewood Cliffs, New Jersey.
10. Howard, Jane. September 10, 1971. "The Canned Menace Called Botulism." *Life.* Vol. 71, No. 11.

BIBLIOGRAPHY

11. Kautter, D. A. and R. K. Lynt, Jr. November 1971. "Botulism." *F.D.A. Papers*. U. S. Department of Health, Education and Welfare. U. S. Printing Office, Washington, D.C.
12. Mallis, Arnold. 1969. *Handbook of Pest Control*. MacNair-Dorland Company, New York.
13. National Academy of Sciences. 1969. *An Evaluation of the Salmonella Problem*.
14. Turner, James S. 1970. *The Chemical Feast*. Grossman Publishers, New York.
15. United States Department of Agriculture. United States Department of Health, Education and Welfare, Public Health Service, Food and Drug Administration. May 1974. GPO 678-871-125/8. U. S. Government Printing Office, Washington, D.C.
16. United States Department of Health, Education and Welfare, Departments of the Army, the Navy and the Air Force. 1969. *Sanitary Food Service*. Instructor's Guide. Public Health Service Publication No. 90. U. S. Government Printing Office, Washington, D.C.
17. Winter, Ruth. 1971. *Beware of the Food You Eat*. Crown Publishers, Inc., New York.
18. Winter, Ruth. 1969. *Poisons in Your Food*. Crown Publishers, Inc., New York.

CHAPTER 2

1. Animal and Plant Health Inspection Service. June 1973. *Salmonella Fact Sheet*. Department of Agriculture.
2. Anonymous. 1971. *Safeguard Your Food*. Good Reading Rack Service, Stamford, Connecticut.
3. Homer, Madean. June 1972. "Safe Handling of Foods in the Home." *F.D.A. Papers*. U. S. Government Printing Office, Washington, D.C.
4. Tannahill, Reay. 1973. *Food in History*. Stein and Day, New York.
5. United States Department of Agriculture. 1965. *Consumers All, The Yearbook of Agriculture*. U. S. Government Printing Office, Washington, D.C.
6. United States Department of Agriculture. March 21, 1973. *Food in the News*.
7. United States Department of Agriculture. April 11, 1973. *Food in the News*.
8. United States Department of Agriculture. April 18, 1973. *Food in the News*.

BIBLIOGRAPHY

9. United States Department of Agriculture. May 30, 1973. *Food in the News.*
10. United States Department of Agriculture. July 11, 1973. *Food in the News.*
11. United States Department of Agriculture. June 1969. *Keeping Food Safe to Eat.* Home and Garden Bulletin No. 162.
12. United States Department of Health, Education and Welfare. *We Want You to Know About Protecting Your Family From Foodborne Illness.* Public Health Service. Food and Drug Administration. DHEW Publication No. (FDA) 74-2003. U. S. Government Printing Office, Washington, D.C.
13. United States Department of Health, Education and Welfare. 1972. *We Want You to Know What We Know About Salmonella.* HEW Publication No. (FDA) 73-2004.

CHAPTER 3

1. Anonymous. 1964. "What You Can Do With Enzymes." *Food Engineering,* Vol. 36, No. 5.
2. De Kruf, Paul. 1926. *The Microbe Hunters.* Harcourt, Brace and Company, New York.
3. Frazier, W. C. 1967. *Food Microbiology.* McGraw-Hill Book Company, New York.
4. Jay, James M. 1970. *Modern Food Microbiology.* Van Nostrand-Reinhold Company, New York.
5. Longree, Karla, Ph.D. 1972. "Quantity Food Sanitation." *Wiley-Interscience.* John Wiley & Sons, Inc., New York.
6. Meyer, Lillian H. 1960. *Food Chemistry.* Reinhold Publishing Corporation, New York.
7. Pelczar, M. J. and R. D. Reid. 1965. *Microbiology.* McGraw-Hill Book Company, New York.
8. Schultz, H. W. 1960. *Food Enzymes.* Avi Publishing Company, Inc., Westport, Connecticut.
9. United States Department of Agriculture. 1966. *Protecting Our Food, The Yearbook of Agriculture.* U. S. Government Printing Office, Washington, D.C.
10. Weiser, H. H., Ph.D.; G. J. Mountney, Ph.D., and W. A. Gould, Ph.D. 1971. *Practical Food Microbiology and Technology.* Avi Publishing Company, Inc., Westport, Connecticut.
11. Wodicka, Virgil O., Ph.D., Director of the Bureau of Foods, Food and Drug Administration. June 26, 1973. "The Food and Drug Administration Cares About Our Food." A speech presented at the 64th Annual Meeting of the American Home Economics Association, Atlantic City, New Jersey.

CHAPTER 4

1. American Meat Institute Foundation. 1960. *The Science of Meat and Meat Products*. W. H. Freeman and Company, San Francisco.
2. Anonymous. November 11, 1973. "Hepatitis Victims All Ate Oysters." *The Denver Post*, Denver, Colorado.
3. Anonymous. November 2, 1973. "Spoiled Meat Found in Markets." *Rocky Mountain News*, Denver, Colorado.
4. Anonymous. November 9, 1973. "Scientist Links Chicken Disease With Cancer." *Rocky Mountain News*, Denver, Colorado.
5. Board, R. G.; J. C. Ayres, A. A. Kraft, and R. H. Forsythe. 1964. "The Microbiological Contamination of Egg Shells and Egg Packing Materials." *Poultry Science*, No. 43.
6. Borgstrom, Georg. 1969. "Principles of Food Science." Volume 1. *Food Technology*. The Macmillan Company, Collier-Macmillan Limited, London.
7. Cruess, W. V. 1958. *Commercial Fruit and Vegetable Products*. 4th ed. McGraw-Hill Book Company, New York.
8. Frazier, W. C. 1967. *Food Microbiology*. McGraw-Hill Book Company, New York.
9. Matz, S. A. 1960. *Baking Technology and Engineering*. Avi Publishing Company, Inc., Westport, Connecticut.
10. Meyer, Lillian H. 1968. *Food Chemistry*. Reinhold Publishing Corporation, New York.
11. Nagel, C. W.; K. L. Simpson, R. H. Vaughn, and G. F. Stewart. 1960. "Microorganisms Associated With Spoilage of Refrigerated Poultry." *Food Technology*, No. 14.
12. Reay, G. A. and J. M. Shewan. 1949. "The Spoilage of Fish and Its Preservation by Chilling." *Advances Food Research*, No. 2.
13. Thatcher, F. S. and J. Montford. 1962. "Egg-Products as a Source of Salmonellae in Processed Foods." *Canadian Journal of Public Health*, No. 53.
14. Toffler, Alvin. 1970. *Future Shock*. Random House, New York.
15. Universal Foods Corporation. Home Service Release No. 3. 433 East Michigan Street, Milwaukee, Wisconsin 53221.
16. United States Department of Agriculture. January 10, 1973. *Food in the News*. U. S. Government Printing Office, Washington, D.C.
17. United States Department of Agriculture. March 21, 1973. *Food in the News*. U. S. Government Printing Office.
18. United States Department of Agriculture. April 1973. *Food in the News*. U. S. Government Printing Office.

BIBLIOGRAPHY

19. United States Department of Agriculture. 1971. *Nutrition: Food at Work for You.* Reprint from Home and Garden Bulletin No. 1, *Family Fare.* Stock No. 0100-1517. U. S. Government Printing Office.
20. United States Department of Agriculture. April 1972. *Toward the New. A Report on Better Foods and Nutrition From Agricultural Research.* Agricultural Information Bulletin No. 341. Stock No. 0100-1565. U. S. Government Printing Office.
21. Wheat Flour Institute. 1971. *From Flour to Bread.* 14 East Jackson Boulevard, Chicago, Illinois 60604.
22. Yeutter, Clayton. September 24, 1973. *Meat and Poultry Inspection in Changing Times.* U. S. Department of Agriculture Press Release. 2911-73 (4687).

CHAPTER 5

1. Anonymous. *Let's Clean House.* Procter & Gamble Co., P.O. Box 14009, Cincinnati, Ohio 45214.
2. Balderston, Lydia Ray. 1935. *Housekeeping Workbook—How to Do It.* J. B. Lippincott Co., Chicago, Philadelphia.
3. Emily Griffith Opportunity School, Denver Public Schools. *Home Care Short Cuts.* Adult, Vocational and Technical Education Division, 1250 Welton Street, Denver, Colorado.
4. General Electric Company. *Quick Shopping Tips—Portable Appliances.* Home Economics Housewares—Business Dept. Vol. No. HE 6422. 1285 Boston Avenue, Bridgeport, Connecticut 06602.
5. Harrison, Molly. 1972. *The Kitchen in History.* Ospry, England.
6. Levitt, Benjamin. 1967. *Oils, Detergents and Maintenance Specialties.* Chemical Publishing Co., Inc., New York.
7. MacLeod, Sarah J. 1915. *The Housekeeper's Handbook of Cleaning.* Harper & Brothers, Publishers, New York, London.
8. *Minneapolis Star.* November 30, 1972. "Flu-Like Shigellosis Is Raging in State." Minneapolis, Minnesota.
9. Richards, Ellen H. 1907. *Sanitation in Daily Life.* Whitcomb & Barrows, Boston.
10. Richards, Ellen H. and S. Marice Elliott. 1910. *The Chemistry of Cooking and Cleaning.* Whitcomb and Barrows, Boston.
11. The Soap and Detergent Association. 1973. *Understanding Automatic Dishwashing.* 475 Park Avenue South, New York, New York 10016.
12. White, James C., Ph.D., Professor of Food Science, New York State College of Agriculture and Life Sciences, Cornell University. May 1972. "Bacterial Contamination of Food." *The Cornell H.R.A. Quarterly,* Cornell University.

BIBLIOGRAPHY

CHAPTER 6

1. Balderston, L. R. 1935. *Housekeeping Workbook—How to Do It*. J. B. Lippincott Co., Chicago, Philadelphia.
2. Busvine, James R., Ph.D., D.Sc., F.I. BIOL. 1966. *Insects and Hygiene*. Methuen and Co., Ltd., London.
3. *Denver Post*. November 14, 1973. "DHA 'Exporting' Cockroaches."
4. Greenberg, B. 1967. "Flies and Diseases." *Scientific American*, 213 (1): 92–99.
5. Herrick, Glenn W. 1914. *Insects Injurious to the Household and Annoying to Man*. The Macmillan Company, New York.
6. McMillan, Wheeler. 1965. *Bugs or People*. Appleton-Century, New York.
7. MacLeod, Sarah J. 1915. *The Housekeeper's Handbook of Cleaning*. Harper and Brothers, Publishers, New York, London.
8. Mallis, Arnold. 1969. *Handbook of Pest Control*. MacNair-Dorland Co., New York.
9. Mampe, Dr. C. Douglass. December 1972. "The Relative Importance of Household Insects in the Continental United States." *Pest Control*, Vol. 40, No. 12.
10. Ong, E. R. de. 1960. *Chemical and Natural Control of Pests*. Reinhold Publishing Corporation, New York.
11. Richards, Ellen H. and S. M. Elliott. 1910. *The Chemistry of Cooking and Cleaning*. Whitcomb and Barrows, Boston.
12. United States Department of Agriculture. 1971. *Controlling Household Pests*. Home and Garden Bulletin No. 96. U. S. Government Printing Office, Washington, D.C.
13. United States Department of Agriculture. June 1969. *Keeping Food Safe to Eat*. Home and Garden Bulletin No. 162. U. S. Government Printing Office.
14. United States Department of Agriculture. September 1970. *Protecting Home-cured Meat From Insects*. Home and Garden Bulletin No. 109. U. S. Government Printing Office.
15. Watt, J. and D. R. Lindsay. 1948. "Effect of Fly Control in a High Morbidity Area." Diarrheal Disease Control Studies. U. S. Public Health Service Reports, Vol. 63, No. 41.

CHAPTER 7

1. Anonymous. March 1969. *Drying Foods at Home*. Circular No. 289. University of Arizona Cooperative Extension Service, Safford, Arizona.
2. Anonymous. 1969. *Eat to Live*. Wheat Flour Institute. 14 East Jackson Boulevard, Chicago, Illinois 60604.

BIBLIOGRAPHY

3. Anonymous. 1972. *Kerr Home Canning Book and How to Freeze Foods.* Kerr Glass Manufacturing Corporation. Sand Springs, Oklahoma.
4. Anonymous. *Knowing Your Home Appliances.* Home Service Center. Public Service Company, Denver, Colorado.
5. Anonymous. 1970. *Meat and Poultry Care Tips for You.* Home and Garden Bulletin No. 174. U. S. Government Printing Office, Washington, D.C.
6. Anonymous. 1971. *Safeguard Your Food.* Good Reading Service. Good Reading Communications, Inc., Stamford, Connecticut 06904.
7. Ball Service Department. 1971. *Successful Home Canning.* Ball Corporation. Muncie, Indiana 47302.
8. Borgstrom, Georg. 1969. "Principles of Food Science." Vol. 1. *Food Technology.* The Macmillan Company, London.
9. Frazier, W. C. 1967. *Food Microbiology.* McGraw-Hill, New York.
10. Geller, Sigmund. August 1, 1972. *Botulism.* Service in Action Bulletin No. 9.305. Colorado State University Cooperative Extension Service, Fort Collins, Colorado.
11. Henry, Betty Lou. August 15, 1972. *Canning Low-Acid Vegetables.* Service in Action Bulletin No. 9.306. Colorado State University Cooperative Extension Service, Fort Collins, Colorado.
12. Henry, Betty Lou. August 1, 1972. *Making Pickles at Home.* Service in Action Bulletin No. 9.304. Colorado State University Cooperative Extension Service, Fort Collins, Colorado.
13. Hertzberg, Ruth, Beatrice Vaughan, and Janet Greene. 1973. *Putting Food By.* The Stephen Greene Press. Brattleboro, Vermont.
14. Home Service Center, Rubbermaid, Inc. Winter, 1972–73. *Freezer Know-How.* Rubbermaid, Inc., Wooster, Ohio 44691.
15. Kerr Research and Educational Department. *10 Short Lessons in Canning and Freezing.* Kerr Manufacturing Corporation. Sand Springs, Oklahoma.
16. National Canners Association. 1971. *The Canning Industry.* Communications Services, 1133 20th Street, N.W., Washington, D.C. 20036.
17. Potter, Norman N., Ph.D. 1968. *Food Science.* Avi Publishing Company, Inc., Westport, Connecticut.
18. Residential Marketing Group. 1965. *Home Management Guidebook for Electric Living.* Edison Electric Institute, New York.
19. Tannahill, Reay. 1973. *Food in History.* Stein and Day, New York.
20. United States Department of Agriculture. January 20, 1974. *Consumer Spots.* U. S. Government Printing Office.

BIBLIOGRAPHY

21. United States Department of Agriculture. December 10, 1973. *Food and Home Notes.* U. S. Government Printing Office.
22. United States Department of Agriculture. July 11, 1973. *Food in the News.* U. S. Government Printing Office.
23. United States Department of Agriculture. September 12, 1973. *Food in the News.* U. S. Government Printing Office.
24. United States Department of Agriculture. December 19, 1973. *Food in the News.* U. S. Government Printing Office.
25. United States Department of Agriculture. 1972. *Home Canning of Meat and Poultry.* Home and Garden Bulletin No. 106. U. S. Government Printing Office.
26. United States Department of Agriculture. 1967. *Home Care of Purchased Frozen Foods.* Home and Garden Bulletin No. 69. U. S. Government Printing Office.
27. United States Department of Agriculture. 1971. *Keeping Food Safe to Eat.* Home and Garden Bulletin No. 162. U. S. Government Printing Office.
28. United States Department of Agriculture. 1966. *Protecting Our Food.* The Yearbook of Agriculture. U. S. Government Printing Office.
29. Winter, Ruth. 1972. *A Consumer's Dictionary of Food Additives.* Crown Publishers, Inc., New York.

CHAPTER 8

1. Anonymous. *101 Meat Cuts.* National Live Stock and Meat Board, 36 South Wabash Avenue, Chicago, Illinois 60603.
2. Anonymous. 1969. *Eat to Live.* Wheat Flour Institute, 14 East Jackson Boulevard, Chicago, Illinois 60604.
3. Anonymous. 1971. *Fresh From the West.* Western Growers Association, 3091 Wilshire Boulevard, Los Angeles, California 90010.
4. Anonymous. *The Time of Your Life.* Pamphlet No. 586525 Rev. B., Sears, Roebuck Company.
5. Armour and Company. 1971. *Meat Guide.* Consumer Service Department, Post Office Box 9222, Chicago, Illinois 60690.
6. Beef Industry Council. 1972. *A Steer Is Not All Steak.* National Live Stock and Meat Board, 36 South Wabash Avenue, Chicago, Illinois 60603.
7. Communications Division. *Watch the Mark of Zero!* American Frozen Food Institute, One Illinois Center, 111 East Wacker Drive, Chicago, Illinois 60601.
8. Dow Chemical Company. 1965. *Guides to Goodness From Your Home Freezer.* Midland, Michigan.
9. General Mills, Inc. 1973. *Your Food Dollar.* A General Mills Consumer Publication. Minneapolis, Minnesota 55440.

BIBLIOGRAPHY

10. Hertzberg, Ruth, Beatrice Vaughan and Janet Greene. 1973. *Putting Food By*. The Stephen Greene Press, Brattleboro, Vermont 05301.
11. Home Service Committee. 1968. *Facts About Food Freezing*. Edison Electric Institute, New York.
12. Lynch, Jeannette, Consumer Marketing Specialist. June 1967. *Buy Times for Fruits and Vegetables*. Pamphlet No. 50. Cooperative Extension Service. Colorado State University, Fort Collins, Colorado.
13. National Canners Association. 1971. *The Canning Industry*. Communication Services, 1133 20th Street, N.W., Washington, D.C. 20036.
14. Office of Consumer Affairs. August 15, 1973. *Consumer News*. U. S. Government Printing Office, Washington, D.C.
15. The Canning Trade. 1969. *A Complete Course in Canning*. Revised and enlarged by Anthony Lopes, Ph.D., 2619 Maryland Avenue, Baltimore, Maryland 21218.
16. Time-Life Books. 1971. *Kitchen Guide*. New York.
17. United States Department of Agriculture. 1968. *Cereals and Pasta in Family Meals*. Home and Garden Bulletin No. 150. U. S. Government Printing Office.
18. United States Department of Agriculture. 1965. "Consumers All." *The Yearbook of Agriculture*. U. S. Government Printing Office.
19. United States Department of Agriculture. February 4, 1974. *Food and Home Notes*. U. S. Government Printing Office.
20. United States Department of Agriculture. 1969. "Food for Us All." *Yearbook of Agriculture*. U. S. Government Printing Office.
21. United States Department of Agriculture. April 11, 1973. *Food in the News*. U. S. Government Printing Office.
22. United States Department of Agriculture. January 16, 1974. *Food in the News*. U. S. Government Printing Office.
23. United States Department of Agriculture. 1973. *Freezing Combination Main Dishes*. Home and Garden Bulletin No. 40. U. S. Government Printing Office.
24. United States Department of Agriculture. 1967. *Home Canning of Fruits and Vegetables*. Home and Garden Bulletin No. 8. U. S. Government Printing Office.
25. United States Department of Agriculture. 1973. *Home Care of Purchased Frozen Foods*. Home and Garden Bulletin No. 69. U. S. Government Printing Office.
26. United States Department of Agriculture. 1973. *Home Freezing of Fruits and Vegetables*. Home and Garden Bulletin No. 10. U. S. Government Printing Office.

27. United States Department of Agriculture. 1970. *Home Freezing of Poultry*. Home and Garden Bulletin No. 70. U. S. Government Printing Office.
28. United States Department of Agriculture. 1971. *How to Buy Canned and Frozen Fruits*. Home and Garden Bulletin No. 191. U. S. Government Printing Office.
29. United States Department of Agriculture. 1969. *How to Buy Canned and Frozen Vegetables*. Home and Garden Bulletin No. 167. U. S. Government Printing Office.
30. United States Department of Agriculture. 1971. *How to Buy Cheese*. Home and Garden Bulletin No. 193. U. S. Government Printing Office.
31. United States Department of Agriculture. 1972. *How to Buy Dairy Products*. Home and Garden Bulletin No. 201. U. S. Government Printing Office.
32. United States Department of Agriculture. 1970. *How to Buy Dry Beans, Peas and Lentils*. Home and Garden Bulletin No. 177. U. S. Government Printing Office.
33. United States Department of Agriculture. 1968. *How to Buy Eggs*. Home and Garden Bulletin No. 144. U. S. Government Printing Office.
34. United States Department of Agriculture. 1967. *How to Buy Fresh Fruits*. Home and Garden Bulletin No. 141. U. S. Government Printing Office.
35. United States Department of Agriculture. 1967. *How to Buy Fresh Vegetables*. Home and Garden Bulletin No. 143. U. S. Government Printing Office.
36. United States Department of Agriculture. 1967. *How to Buy Instant Nonfat Dry Milk*. Home and Garden Bulletin No. 140. U. S. Government Printing Office.
37. United States Department of Agriculture. 1971. *How to Buy Lamb*. Home and Garden Bulletin No. 195. U. S. Government Printing Office.
38. United States Department of Agriculture. 1972. *How to Buy Meat for Your Freezer*. Home and Garden Bulletin No. 166. U. S. Government Printing Office.
39. United States Department of Agriculture. 1972. *How to Buy Potatoes*. Home and Garden Bulletin No. 198. U. S. Government Printing Office.
40. United States Department of Agriculture. 1968. *How to Buy Poultry*. Home and Garden Bulletin No. 157. U. S. Government Printing Office.

BIBLIOGRAPHY

41. United States Department of Agriculture. 1967. *How to Use USDA Grades in Buying Food.* Home and Garden Bulletin No. 196. U. S. Government Printing Office.
42. United States Department of Agriculture. 1972. *Keeping Food Safe to Eat.* Home and Garden Bulletin No. 162. U. S. Government Printing Office.
43. United States Department of Agriculture. 1972. *Meat & Poultry Care Tips for You.* Home and Garden Bulletin No. 174. U. S. Government Printing Office.
44. United States Department of Agriculture. 1969. *Meat & Poultry Standards for You.* Home and Garden Bulletin No. 171. U. S. Government Printing Office.
45. United States Department of Agriculture. 1967. *Milk in Family Meals.* Home and Garden Bulletin No. 127. U. S. Government Printing Office.
46. United States Department of Agriculture. 1973. *Storing Perishable Foods in the Home.* Home and Garden Bulletin No. 78. U. S. Government Printing Office.
47. United States Department of Agriculture. 1972. *Toward the New.* Agriculture Informational Bulletin No. 341. U. S. Government Printing Office.
48. United States Department of Agriculture. 1973. *Your Money's Worth in Foods.* Home and Garden Bulletin No. 183. U. S. Government Printing Office.
49. United States Department of Agriculture. 1967. *What to Do When Your Home Freezer Stops.* Leaflet No. 321. U. S. Government Printing Office.
50. United States Department of Health, Education and Welfare. 1968. *Cold Facts About Home Food Protection.* Public Health Service Publication No. 1247. U. S. Government Printing Office.
51. United States Department of Health, Education and Welfare, Public Health Service, Consumer Protection and Environmental Health Service. 1969. *Sanitary Food Service, Instructor's Guide.* Public Health Service Publication 90. U. S. Government Printing Office.
52. United States Department of Health, Education and Welfare, Public Health Service, Food and Drug Administration. 1971. *Some Questions and Answers About Canned Foods.* FDA Fact Sheet No. CSS F8-1-71. U. S. Government Printing Office.
53. United States Department of Health, Education and Welfare, Public Health Service, Food and Drug Administration. 1974. *We Want You to Know About Protecting Your Family From Foodborne Illness.* DHEW Publication No. (FDA) 74-2003. U. S. Government Printing Office.

BIBLIOGRAPHY

54. University of Arizona Cooperative Extension Service. 1969. *Drying Foods at Home*. Circular No. 289. University of Arizona, College of Agriculture, Tucson, Arizona.
55. White, Philip L., Sc.D., Editor and Director, Department of Foods and Nutrition; Secretary, Council on Foods and Nutrition, American Medical Association. 1970. *Let's Talk About Food*. Division of Scientific Activities, American Medical Association Circulation and Records Department, 535 North Dearborn Street, Chicago, Illinois 60610.

CHAPTER 9

1. Anonymous. July 1971. "How Safe Is the Food You Buy?" *Good Housekeeping*.
2. Anonymous. February 22, 1974. "New FDA Rules for Food Processing." *Medical World*.
3. Anonymous. March 1973. "The High-Filth Diet, Compliments of FDA." *Consumer Reports*.
4. Anonymous. September 1971. "When Americans Are a Swallow Away From Death." *Today's Health*.
5. Anonymous. February 1972. "You Can Reduce the Risks of Food Poisoning." *Good Housekeeping*.
6. Borgeson, Lillian. June 1972. "How Much Danger in Canned Foods?" Condensed from *Family Health*. *Reader's Digest*.
7. Committee on Interstate and Foreign Commerce, United States House of Representatives and its Subcommittee on Public Health and Environment. January 1974. *A Brief Legislative History of the Food, Drug and Cosmetic Act*. U. S. Government Printing Office, Washington, D.C.
8. Frazier, W. C. 1967. *Food Microbiology*. McGraw-Hill Company, New York, London.
9. Kennedy, Ted, Senator (D-Mass.). December 7, 1973. *Food Amendments of 1973*. Report No. 93-605. United States Senate Committee on Labor and Public Welfare. U. S. Government Printing Office.
10. Minneapolis *Star*. October 12, 1972. "128 Found Ailing After Picnic, Buffet."
11. Nader, Ralph. January 1973. "Ralph Nader Reports." *Ladies' Home Journal*.
12. Ross, Irwin. September 1972. "How Safe Is Our Food?" *Reader's Digest*.
13. United States Department of Agriculture. 1969. *Food for Us All*. The Yearbook of Agriculture. U. S. Government Printing Office.

BIBLIOGRAPHY

14. United States Department of Agriculture. October 1972. *Foreign Meat and Poultry Inspection Program*. Animal and Plant Health Inspection Service. U. S. Government Printing Office.
15. United States Department of Agriculture. February 1972. *Meat and Poultry, Care Tips for You*. Home and Garden Bulletin No. 174. U. S. Government Printing Office.
16. United States Department of Agriculture. December 1969. *Meat and Poultry, Clean for You*. Home and Garden Bulletin No. 173. U. S. Government Printing Office.
17. United States Department of Agriculture. October 1969. *Meat and Poultry Standards for You*. Home and Garden Bulletin No. 171. U. S. Government Printing Office.
18. United States Department of Agriculture. October 1969. *Meat and Poultry, Wholesome for You*. Home and Garden Bulletin No. 170. U. S. Government Printing Office.
19. United States Department of Agriculture. August 14, 1973. *News*. USDA Fact Sheet No. 2507-73. U. S. Government Printing Office.
20. United States Department of Agriculture. 1966. *Protecting Our Food. The Yearbook of Agriculture*. U. S. Government Printing Office.
21. United States Department of Health, Education and Welfare. February 1, 1974. *Consumer News*. DHEW Publication No. (05) 74-108. U. S. Government Printing Office.
22. United States Department of Health, Education and Welfare, Public Health Service, Food and Drug Administration. August 1972. *Federal Food, Drug and Cosmetic Act as Amended*. U. S. Government Printing Office.
23. United States Department of Health, Education and Welfare, Public Health Service, Food and Drug Administration. 1973. *How You Can Be an Extension of FDA*. DHEW Publication No. (FDA) 73-1015. U. S. Government Printing Office.
24. United States Department of Health, Education and Welfare, Public Health Service, Food and Drug Administration. 1973. *We Want You to Know What We Know About the Laws Enforced by FDA*. DHEW Publication No. (FDA) 73-1031. U. S. Government Printing Office.
25. Wodicka, Virgil O., Ph.D. March 1972. *The Current Status of Food Regulation*. Reprint from FDA Papers. DHEW Publication No. (FDA) 73-2010. U. S. Government Printing Office.
26. Yeutter, Clayton, Assistant Secretary of Agriculture. September 24, 1973. *Meat and Poultry Inspection in Changing Times*. An address before the American Meat Institute, Chicago, Illinois. United States Department of Agriculture News Release. 2911-73. U. S. Government Printing Office.

Index

Acarus siro, 132–33
Aerosol sprays, 136–37
Air, as source of contamination, 49–50
Alcohol, as preservative, 167
Alternaria, 39
Aluminum cooking utensils, cleaning, 108
Amanita muscaria, 16
Animals, as source of contamination, 47–48
Ants, 121–23, 138
Apples, 68, 182, 208
Appliances. *See* Kitchen appliances
Apricots, 68, 182, 208
Artichokes, 186, 209
Asepsis in food preservation, 141
Asparagus, 68, 186, 209, 220
Asperquillus, 38
Avocados, storing, 68, 209

Bacteria
  characteristics of, 40–42
  commonly found in food, 42–45
  first discovered, 35
  spoilage caused by, 33
  temperatures controlling, 28, 168
  See also specific bacteria
Baking powder, storing, 213
Bananas, 182, 208
Beans
  buying, 186
  storing, 68, 69, 209, 217–18, 220
Beef. *See* Meats
Beetles, 129, 138
Beets, 68, 186
Berries, 68, 182, 185, 208
Beverages, storing, 214–15, 221
Blenders, cleaning, 109
Blueberries, buying, 182
Boiling-water bath canning method, 144, 147, 168
Botulism, 2–5, 18
Bread boards, cleaning, 108
Bread mold (*Rhizopus*), 38
Breads, 55–59
  freezing home-baked, 156
  safeguards for, 57–59
  spoilage of, 55–57
  storing, 215–16, 220, 221
Brettanomyces, 40
Brisket (beef cut), described, 173–74

INDEX

Broccoli, 68, 186
Brussels sprouts, 187, 210
Butter, storing, 206, 221

Cabbage, 68, 187, 210
Cake mixes, 28, 212
Candida, 40
   lipolytica, 40
Canned fish, storing, 203
Canned meats, storing, 202
Canned milk, storing, 205
Canned poultry, storing, 202
Canned produce
   buying, 190
   storing, 218–19
Canning, 23–24, 28–29, 143–50
Can openers, cleaning, 108
Carrots, 69, 187, 210
Cauliflower, 69, 187, 210
Celery, 69, 187, 210
Cellar storage, principles of, 151–52
Cereals, 55–59
   safeguards for, 57–59
   spoilage of, 55–57
   storing, 216
Cheese
   buying, 179–80
   storing, 206–7
Chemical disinfectants, 96
Chemical poisoning, 13–15
Cherries, 68, 182, 208
Chilling, preservation by, 152–53
Choice (beef grade), 174
Chuck (beef cut), described, 173
Citnobacter cloacae, 74
Citrus fruits
   buying, 182–84
   storing, 208
Clams, buying, 176
Cleanliness, kitchen, 95–113. *See also* Personal hygiene
Clostridium
   in milk products, 81
   perfringens, 8–9, 18
   on sugars, 61
Clostridium botulinum, 2–4
   in canned foods, 139, 143–45
   destroyed by radiation, 167
   in frozen vegetables, 66
   thawing process promoting growth of, 157
Cockroaches, 118–20, 138
Coffeemakers, cleaning, 109
Combination dishes
   freezing, 156
   refreezing, 157
   storing, 220
Commercial (beef grade), 175
Commercial agencies involved in food control, 231
Complaints, reporting, 242–43
Control measures
   for botulism, 5
   for *Clostridium perfringens,* 8–9
   for infectious hepatitis, 13
   for salmonellae, 11–12
   for staphylococcal poisoning, 8
   for trichinosis, 14
Cooked meats, refrigeration of, 24–25
Cooking time for meats and poultry, 22–24
Cooking utensils
   cleaning aluminum, 108
   contamination through, 22
Corn, 69, 187, 210, 220
Cornmeal, storing, 212
Corynebacterium, 43
Coxsackie virus, 13
Crabs, buying, 176
Cranberries, 182, 208
Cucumbers, 68, 187, 210
Cured meats, 28
   storing of, 201–2
   temperature control for, 92

Dairy mold (*Geotrichum*), 38
Dairy products. *See* Butter; Cheese;

INDEX

Milk and milk products
Dates, storing, 208
Debaryomyces kloeckeri, 40
Dishwashers, cleaning, 105–6
Dishwashing, 110
Disinfection of kitchens, 96–98
Dried eggs, 26
 storing, 204
Dried foods, storing, 218
Dried legumes, storing, 217–18
Drosophila melanogaster, 131–32
Drying
 of meats, 90
 preservation by, 161–66
Dry milk, storing, 205
Dusts (insecticides), 137

Echo viral infections, 13
Eggplants, 187, 210
Egg Product Inspection Act, 227–28
Eggs
 buying, 178–79
 freezing, 156
 proper handling of, 25–26
 safeguards for, 73–77
 spoilage of, 72–73
 storing, 204
Electric frying pans, cleaning, 109
Electric knives, cleaning, 109
Electric mixers, cleaning, 109
Enzymes, spoilage caused by, 34, 45–47
Erwinia, 43
Escherichia, 43, 74

Fair Packaging and Labeling Act, 226, 228
Fats, storing, 213–14
FDA. *See* Food and Drug Administration
Federal agencies involved in food control, 228–29
Federal inspections, 231–36
Figs, storing, 208

Fish, 69–72
 buying, 175–76
 canning, 147, 148
 freezing, 155–56
 refreezing, 159, 160
 safeguards for, 71–72
 spoilage of, 70–71
 storing of, 203, 221
 thawing out frozen, 27
Flank (beef cut), described, 174
Flavobacterium, 43
Flour, storing, 212
Food, Drug and Chemical Act, 166
Food, Drug and Cosmetic Act (1938), 225–26, 228
Food Additives Amendment (1958), 166
Food and Drug Administration (FDA)
 chemical poisoning and, 16
 in investigation and control of food, 223–26, 228–36, 238, 242–43
 kitchen hygiene recommendations of, 108
 preservation by radiation approved by, 167
Food controls, 223–44
Food disposers, cleaning, 106–7
Food handling, personal hygiene in, 19–21
Food poisoning, 1–18
 investigating outbreaks of, 239–42
Food shopping, 169–90
Foreshank (beef cut), described, 173
Freezer storage chart, 220–21
Freezing, 88, 153–61, 190, 194–99
Fresh fruits
 buying, 180–85
 storing, 68–69, 207–9, 221
Fresh vegetables
 buying, 185–89
 storing, 68–69, 209–11, 221

INDEX

Frozen eggs, 26
Frozen vegetables
  Clostridium botulinum in, 66
  storage of, 220
Fruit flies (vinegar flies, pomace flies), 131–32
Fruit juices, refreezing, 159
Fruits
  canning of, 144, 147
  canning temperatures for, in waterbath method, 168
  chilling of, 152–53
  drying, 161, 163, 164
  refreezing, 159
  safeguards for, 65–69
  as source of spoilage, 47
  spoilage of, 63–65
  storage of, 68–69
  washing, 21–22
  See also Fresh fruits

Garlic, storing, 210
Geotrichum (dairy mold), 38
Good (beef grade), 174–75
Grading, 237
  beef, 174–75
  canned produce, 190
  cheese, 180
  eggs, 178–79
  fish and seafood, 176
  fresh fruits, 181
  fresh vegetables, 186
  poultry, 177
Grain products, 55–59
  safeguards for, 57–59
  spoilage of, 55–57
  See also specific grain products
Grapefruits, 68, 182–83
Grapes, 68, 183, 208
Green plants, as source of spoilage, 47

Hamburger meat, 28–29
Hams. *See* Cured meats

Heating of frozen dishes, 157–59
High temperatures
  molds destroyed by, 39
  preservation by use of, 142–50
Home-canned foods, detecting spoilage of, 28–29
House flies, 120–21, 138
House pests, 115–38

Ice cream, 160, 206, 220
Ice crushers, cleaning, 109
Ice-cube compartment freezing, 197–99
Import Meat Act, 227, 229
Import Milk Act, 227, 228
Incubation time of various poisonings, 18
Infectious hepatitis, 12–13
Infectious jaundice, transmitted by rats, 126
Insecticides
  for ants, 122–23
  for cockroaches, 118–20
  for house flies, 121

Kitchen appliances
  cleaning large, 98–107
  cleaning portable, 107–9
Kitchen cleanliness, 95–113

Lactobacillus, 43
Lamb. *See* Meats and Meat Products
Leftovers, refrigeration of, 191, 192, 221
Legislation for food control, 225–31
Lemons, 68, 183
Lettuce, 188, 210
Limes, 68, 183
Lobster, buying, 176
Low-acid foods
  canning temperatures for, 168
  Clostridium botulinum in, 143, 145
  refreezing, 159

INDEX

Low temperatures, preservation by use of, 150–61. *See also* Freezing; Refrigeration

Margarine, storing, 206, 221
Mayonnaise-containing foods, 26
Meat grinders, cleaning, 108
Meat Inspection Act, 224, 229
Meats and meat products, 85–93
 buying, 171–75
 canning of, 23–24, 28–29, 147–48
 contamination of, 85
 cooking time for, 22–24
 drying, 161, 165
 freezing, 155
 recommended internal temperatures for cooking, 31–32
 refreezing, 160
 refrigeration of cooked, 24–25
 safeguards for, 87–93
 spoilage of, 86–87
 storing, 201–2, 220, 221
 thawing out, 27
 time needed to thaw, in refrigerator, 222
 See also Poultry
Melons, 68, 183–84, 208–9
Mice, 123–24
Micrococcus, 43
Microorganisms, food preservation through removal of, 141–42
Milk and milk products, 80–84
 safeguards for, 83–85
 spoilage of, 80–83
 storing of, 204–5, 221
Mites, as pantry pests, 132–33
Molds
 characteristics of, 36–38
 common, 38–39
 discovered, 35
 in fish, 71
 spoilage caused by, 33
 temperatures destroying, 28
Mortality rate from various food poisonings, 18
Moths, 130
Mucor, 60
Municipal agencies involved in food control, 230–31
Murine typhus
 mice as transmitters of, 124
 rats as transmitters of, 126
Mushroom poisoning, 16
Mushrooms, 188, 211

Nectarines, 68, 184, 209

Oils, storing of, 213–14
Okra, 188, 211
Onions, 68, 188, 211
Open-kettle canning method, 143, 144
Oranges, 68, 184
Oxalic acid poisoning, 16
Oxidation, defined, 34
Oysters, buying, 176

Paintbrush application of pesticides, 137
Pantry pests, 129–33
Parsnips, 68, 188
Pasteurization, 142–43
Peaches, 68, 184, 209
Pears, 68, 184–85, 209
Peas, 69, 211
Pediococcus, 43, 60
Penicillium, 38
 camemberti, 38
 in eggs, 73
 growth of, on sugars, 60
 roqueforti, 38
Peppers, 69, 188
Perishable foods, basic rule for safe handling of, 28
Personal hygiene
 in food handling, 19–21
 in the kitchen, 111–13
Pest control, 133–38

INDEX

Pesticides
  proper use of, 134–35
  types of, 135–38
  See also Insecticides
Pests, house, 115–38
Pickles
  canning temperatures for, 168
  preservatives in, 167
Pineapples, 69, 185, 209
Plague
  mice as transmitters of, 124
  rats as transmitters of, 125–26
Plate (beef cut), described, 174
Plums, 68, 185, 209
Poisoned baits, 137
  for mice, 124
  for rats, 128–29
Poisonous animals, 16–17
Poisonous plants, 16–17
Poliomyelitis, 13, 126
Pomace flies (vinegar flies, fruit flies), 131–32
Pork. *See* Meats and meat products
Portable kitchen appliances, 107–9
Potatoes, 68, 188, 211
Poultry, 77–80
  buying, 176–78
  canning, 147, 148
  cooking time for, 22–24
  refreezing, 160
  refrigeration of cooked, 24–25
  safeguards for, 78–80
  spoilage of, 77–78
  storing of, 202, 220, 221
  thawing out, 27
  time needed to thaw, in refrigerator, 222
Preservation, 139–68. *See also* Freezing; Refrigeration
Preservatives, preservation with, 166–67
Pressure canning, 144–47, 168
Prime (beef grade), 174

Private agencies involved in food control, 231
Problem foods, safeguards against spoilage of, 53–93
Professional societies involved in food control, 231
Proteus, 43–44
Prunes, 185, 209
Pseudomonas, 44
  fluorescens, 71
"Ptomaine" poisoning. *See* Staphylococcal poisoning
Public Health Service Act, 226
Pure Food and Drug Act (1906), 224

Rabies, rats as transmitters of, 126
Radiation, preservation by, 167
Radishes, 69, 189, 211
Raspberries, 185
Rat-bite fever, 126
Rats, 125–29
Refreezing of thawed foods, 159–60
Refrigeration, 190–94
  of cooked meats and poultry, 24–25
  fundamental rules in, 193–94
  proper methods of, 191–92
Refrigerators, cleaning, 101–4
Refrigerator storage chart, 221
Rhizopus (bread mold), 38
Rhodotorula, 40
Rhubarb, 189
Rib (beef cut), described, 173
Rickettsial pox, rats as transmitters of, 126
Rope disease, 57–58
Round (beef cut), described, 173
Rutabagas, 68, 211

Saccharomyces, 40
Salt
  as preservative, 167
  storing, 213

INDEX

Salmonella, 9–12, 18
  animals as source of, 48
  in canned meats, 24
  characteristics of, 44
  destroyed by cooking, 25
  destroyed by disinfection, 97
  in eggs, 72, 76, 77
  eliminating, 235, 238
  investigating outbreaks of, 240
  major sources of, 20, 22, 124
  pests transmitting, 120
  in poultry, 77–79
  typhimurium, 118
Sarcina, 71
Scallops, 176
Seafoods, 69–72
  buying, 175–76
  canning, 147
  freezing, 155–56
  refreezing, 160
  safeguards for, 71–72
  spoilage of, 70–71
  storing, 203, 221
Sealing of frozen foods, 156–57
Sewage, as source of contamination, 48
Shigella, 12, 44, 240
  flexneri, 1
Shigellosis, 12
Short loin (beef cut), described, 173
Shrimp, buying, 176
Sirloin (beef cut), described, 173
Snakeroot poisoning, 16
Soil, as source of contamination, 48
Space sprays, 136–37
Spices, storing, 214
Spinach, storing, 211, 220
Spoilage, 28–29, 33–93
Sporotrichum, 38, 73
Squash, 68, 69, 189, 211
Standard (beef grade), 175
Staphylococcal poisoning, 1, 5–8, 18
  from hams, 202
  investigating outbreaks of, 240

Staphylococcus, 44, 92
  aureus, 7, 18, 152
  destroyed by disinfection, 97
  thawing promoting growth of, 157
Staples, buying, 189–90
State agencies involved in food control, 230
Storage, 199–222
  guide for fruit and vegetable, 68–69
  principles of cellar, 151–52
  See also Freezing; Refrigeration
Stoves, cleaning, 99–101
Strawberries, buying, 185
Streptococcus, 44, 97
Streptomyces, 44–45, 71, 74
Sugars
  as preservatives, 167
  safeguards for, 61–62
  spoilage of, 59–61
  storing of, 212–13, 216–17
Sunlight, as natural disinfectant, 96
Surfaces, cleaning, 110–11
Sweet potatoes, 68, 189, 211
Symptoms of food poisoning, 18

Tangerines, 185
Tea Act, 228
Tea Importation Act, 226–27
Temperature control
  danger zone in, 27
  for destroying bacteria, 28, 168
  for handling meats, 89–90
  preservation by use of, 150–61
  See also Freezing; High temperatures; Refrigeration
Termites, 138
Thawing, 26–27, 157–59
Toaster-ovens, cleaning, 109
Toasters, cleaning, 109
Tomatoes, 68, 189, 211
Traps
  mouse, 124
  rat, 128–29

INDEX

Trichinella spiralis, 13, 18
Trichinosis, 13–14, 18, 126
Trichothecium, 38
Turnips, 69, 189, 211

Uncooked meats, 220
United States Department of Agriculture (USDA), 2
 canned produce grading by, 190
 canning methods approved by, 144, 145
 cheese grading by, 180
 control of trichinosis and, 14
 egg grading by, 178–79
 fresh vegetable grading by, 181, 186
 meat grading by, 172, 174
 meat spoilage and, 88–89
 poultry grading by, 177
 Salmonella and, 9, 77, 78

Variety meats, 201
Veal. *See* Meats and meat products
Vegetables
 canning, 28–29, 143, 144, 146–47
 chilling, 152–53
 drying, 161–63, 165
 as prime target of *Clostridium botulinum,* 66, 144
 refreezing, 160
 safeguards for, 65–69
 spoilage of, 63–65
 storage of, 68–69
 washing, 21–22
 See also Canned produce; Fresh vegetables
Venison, 23, 221
Vinegar, as preservative, 167
Vinegar flies (fruit flies, pomace flies), 131–32
Viral food infections, 12–13

Waste compactors, cleaning, 107
Water
 purifying, 30–31
 as source of contamination, 48–49
Watermelons, 69
Weevils, as pantry pests, 130–31
Wholesome Meat Act, 227
Wholesome Poultry Products Inspection Act, 227

Yeasts
 characteristics of, 39–40
 common, 40
 discovered, 35
 spoilage caused by, 33
 storing, 217
 temperatures destroying, 28

Zygosaccharomyces, 40